NASA AND THE ENVIRONMENT

THE CASE OF OZONE DEPLETION

by W. Henry Lambright

Monographs in Aerospace History No. 38

NASA SP-2005-4538

National Aeronautics and
Space Administration

NASA History Division
Office of External Relations

NASA Headquarters
Washington, DC

For sale by the Superintendent of Documents, U.S. Government Printing Office
Internet: bookstore.gpo.gov Phone: toll free (866) 512-1800; DC area (202) 512-1800
Fax: (202) 512-2250 Mail: Stop SSOP, Washington, DC 20402-0001

ISBN 0-16-074946-8

Library of Congress Cataloging-in-Publication Data
Lambright, W. Henry, 1939-
NASA and the environment : the case of ozone depletion / by W. Henry Lambright.
p. cm.
Includes bibliographical references.
1. Ozone layer depletion—Environmental aspects—United States. 2. Ozone layer depletion—Government policy—United States. 3. Air—Pollution—Environmental aspects—United States. 4. Air—Pollution—Government policy—United States. 5. Environmental policy—United States. I. Title.

QC879.73.C5L36 2005
363.738'75—dc22   2005001712

Contents

Introduction 1

Stage 1 3

Stage 2 5

Stage 3 7

Stage 4 9

Stage 5 17

Stage 6 25

Stage 7 35

Stage 8 41

Conclusion 47

Endnotes 49

About the Author 59

NASA History Monographs 61

Index 65

Introduction

While the National Aeronautics and Space Administration (NASA) is widely perceived as a space agency, since its inception NASA has had a mission dedicated to the home planet. Initially, this mission involved using space to better observe and predict weather and to enable worldwide communication. Meteorological and communication satellites showed the value of space for earthly endeavors in the 1960s. In 1972, NASA launched Landsat, and the era of earth-resource monitoring began.[1]

At the same time, in the late 1960s and early 1970s, the environmental movement swept throughout the United Sates and most industrialized countries. The first Earth Day event took place in 1970, and the government generally began to pay much more attention to issues of environmental quality. Mitigating pollution became an overriding objective for many agencies. NASA's existing mission to observe planet Earth was augmented in these years and directed more toward environmental quality. In the 1980s, NASA sought to plan and establish a new environmental effort that eventuated in the 1990s with the Earth Observing System (EOS). The Agency was able to make its initial mark via atmospheric monitoring, specifically ozone depletion.

An important policy stimulus in many respects, ozone depletion spawned the Montreal Protocol of 1987 (the most significant international environmental treaty then in existence). It also was an issue critical to NASA's history that served as a bridge linking NASA's weather and land-resource satellites to NASA's concern for the global changes affecting the home planet. Significantly, as a global environmental problem, ozone depletion underscored the importance of NASA's ability to observe Earth from space. Moreover, the NASA management team's ability to apply large-scale research efforts and mobilize the talents of other agencies and the private sector illuminated its role as a "lead" agency capable of crossing organizational boundaries as well as the science-policy divide.

APPROACH

In the analysis below, the approach used to examine the evolving relationship between an agency and a program focuses on decision-making. The decision-making process goes through a number of stages that can span many years.

Stage 1—Awareness. The first stage entails the emergence of a problem that needs public and government attention. Initially, there may be little activity by an agency with respect to the issue. No one is responsible for dealing with the problem at this point.

Stage 2—Trigger. Subsequently, some event occurs that triggers action. Government places the issue on its agenda. It is "framed" as a particular kind of problem or opportunity. Who is in charge of the issue or problem still remains unclear, and there may be numerous parties contesting for ownership.

Stage 3—Establishing a Program. Next, a decision is reached by appropriate government authorities to assign jurisdiction over the issue to an agency. Legislation is passed, which confers legitimacy and resources. The agency establishes a program to cope with the problem. In the case at hand, it is a research and development (R&D) program that is established.

Stage 4—Early Implementation. The agency plans, organizes, and executes a program of action. This stage can involve numerous substages. The nature of the program can change as progress is made, as can the organization.

Stage 5—Evaluation/Reorientation. At some point along the way, there is a pause and the program is evaluated. The evaluation can be formal or informal, scientific or political. The results of the evaluation may lead to various outcomes: a decision to continue the program as is, a plan to reorient it, or a decision to terminate it.

Stage 6—Amplification. This important stage is often overlooked in decision-making literature, perhaps because it does not always occur. When it does occur, it involves the expansion of the program into new areas, even as the existing program continues to be implemented. In other words, not only does the agency reorient the program, but the agency itself changes.

Stage 7—Later Implementation. In this stage, the agency reimplements the program, in a greatly modified organizational and policy context. Scientific progress is made, but so are mistakes. The agency, perhaps carried along by the momentum of stages 5 and 6, overreaches and has to lower the program's profile.

Stage 8—Institutionalization. In the eighth stage, the issue dims in the public's perception and may even cease to be considered as a problem. The program becomes a routine, ongoing agency activity, but it is now one that operates at a lower priority than before.

The above decision-making model has a linear structure that does not exist in reality. However, it conveys, in a general way, the overall course of the decisions being made over time. The ozone depletion issue has moved through the first seven stages at NASA, and it is now advancing into the eighth. The ozone decision-making process began in the late 1960s. Along the way, NASA assumed a new role and developed new relationships with other agencies. It made key decisions in the program's birth and development. While no longer acute, the ozone policy process continues, and there is increased scientific recognition of the link between ozone depletion and climate change. That link, along with other issues remaining to be understood, has required constant attention. Ozone depletion thus represents an important case study in the history of NASA and environmental sciences. It is one from which many lessons can be learned about the management of science and technology and the application of knowledge to policy-making decisions.

In tracking NASA's decision-making process, the author has made use of the various books on ozone policy—and it should be emphasized this paper's orientation is on ozone policy and the NASA government program, not the history of environmental science. For other approaches, see the works of Benedick, Andersen and Sarma, Christie, and Parson.[2] Benedick deals with the diplomatic story; Andersen and Sarma, the role of the United Nations; Christie, the scientific debate and consensus-building process in science; and Parson, most comprehensively, government policy and the evolution of scientific thought. None of these books focuses on NASA.

Stage 1—Awareness

When ozone depletion first became an issue in the late 1960s, NASA was preoccupied with Apollo and landing a man on the Moon. The Department of Transportation (DOT) had primary responsibility for another huge project—the Supersonic Transport (SST). The SST was, at this time, still in development, but it was already a target of the environmental movement, whose power rose rapidly in the latter part of the decade.

Environmentalists attacked the SST for its sonic boom (noise pollution). They also said it would pollute the atmosphere and dilute the ozone layer, which lay in the upper reaches of the atmosphere—a region approximately 15–50 km above Earth's surface known as the stratosphere. If the ozone layer were depleted, dangerous ultraviolet (UV) rays would reach Earth, increasing the incidence of human skin cancer, and affect other living creatures. The Massachusetts Institute of Technology (MIT) conducted a study in 1969 concluding that the SST might well have some impact on the stratosphere, and it recommended that a permanent stratospheric monitoring program be initiated to assess the SST's true impact.[3]

Such a monitoring program did not materialize initially. Instead, there was debate within the scientific community. One scientist who took up the environmentalists' cause was James McDonald of the University of Arizona. He testified before Congress and championed the issue. Because he was also a researcher of unidentified flying objects (UFOs) and extraterrestrial visitation, his credibility was easily attacked. Harold Johnston, a University of California at Berkeley scientist, also concluded SSTs were a possible menace to the ozone layer. A draft of an article he sent to *Science* magazine was leaked to the *New York Times* in 1971. It charged that a fleet of SSTs would deplete the ozone layer enough to allow sufficient radiation to reach Earth's surface and to cause blindness.[4] The National Research Council (NRC) looked into the matter and reported that there was adequate evidence of a potential problem to justify further research.

Senator Clinton Anderson (Democrat, NM), Chair of the Senate Committee on Aeronautical and Space Sciences, believed NASA, the agency he oversaw, had the best expertise to conduct a research program. He wrote to James Fletcher, the NASA administrator at that time, and urged him to take the initiative.[5]

Whatever Fletcher may have thought, NASA was unable to get strongly involved. The congressional debate over SST came to a head in 1971 and Congress terminated the program, much to the dismay of the Nixon administration. All that could be salvaged, as far as DOT was concerned, was a four-year research effort called the Climate Impact Assessment Program (CIAP). It is noteworthy that the program's tasks included looking at SST's impact on climate and the ozone layer.

NASA was largely a bystander to these significant policy events. It worked with DOT on aeronautical research. As discussion of the SST issue increasingly turned to questions of environmental impact, NASA was drawn into these discussions to provide input on the research aspect. However, DOT was still the agency in charge.[6]

Stage 2—Trigger

As far as NASA was concerned, the catalyst for policy action involved internal and external matters. Indeed, it could be said that there were two triggers. The first was the approval by President Nixon of the development of a Space Shuttle in 1972. Given the demise of SST the year before, NASA leadership understood that potential opposition to the Space Shuttle could result in the Agency being charged with depleting the ozone layer or affecting climate via the Shuttle. A report commissioned by NASA in 1973 found that the Space Shuttle would release chlorine, a highly reactive element theorized to destroy ozone in the stratosphere.[7] The report alarmed NASA management, and Johnson Space Center's (Johnson) initial response was to suppress the information. However, NASA Headquarters overruled Johnson. As soon as this stance was reversed, the report was published, and an office was established to study the Shuttle's environmental effects.

In 1974, NASA also sponsored a workshop that explored the pollution issues further. Fletcher decided that NASA could not rely on DOT's CIAP—which was going to end in the near term. It had to involve itself in stratospheric research in a much more serious, proactive way. These matters were related directly to NASA's central mission and its dominant program, the Space Shuttle. They required much greater study than they had previously received, and Fletcher ordered officials in the Agency to direct their attention accordingly.[8]

The other trigger had nothing to do with the SST, much less the Space Shuttle. In mid-1974, Mario Molina and F. Sherwood Rowland published a paper in *Nature* magazine in which they suggested that a common family of industrially produced compounds known as chlorofluorocarbons (CFCs) could lead to stratospheric ozone depletion.[9] These were ubiquitous, a clear and present danger, rather than a futuristic threat. They were found in everyday items such as spray cans, air conditioners, refrigerators, and the like. If the ozone layer were to be depleted, the Sun's ultraviolet rays would seep through this protective shield and cause enhanced rates of skin cancer in humans.

The media and environmentalists seized on the theory and called for action. The political conflict, which had cooled since SST's cancellation, reheated rapidly. Congress debated what to do. The options included creating a long-term research program focused on the stratosphere and/or regulating CFCs.[10]

Stage 3—Establishing a New Program

It was much easier to reach a bipartisan political consensus on a new research program than on new regulation. But who should be in charge? From the standpoint of existing missions, the stratosphere was a no-man's land. It was not, technically, "space." Nor was it an area where the National Oceanic and Atmospheric Administration (NOAA) typically operated. DOT was a potential candidate, thus extending CIAP, and the National Science Foundation (NSF) was also a possible lead agency. Another option was to coordinate an interagency program through the sub-cabinet-level Interagency Committee on Atmospheric Sciences (ICAS), a body of the Federal Council on Science and Technology (FCST), headed by the White House Science Adviser.

President Ford's Science Advisor, H. Guyford Stever, wanted a thorough airing of the issues through the interagency FCST before reaching a final conclusion. However, forces in Congress were anxious to move faster. It was obvious that an important new research mission was coming into being. Agencies saw bureaucratic interests at stake.

In late 1974, the "lead agency" question was discussed by John Naugle, Deputy Associate Administrator, NASA, and Ed Todd, Deputy Assistant Director for Research, NSF. In a memo written for the record by Naugle, Todd noted that

> DOT, as a result of the CIAP, probably had the most experience in handling a problem of this nature. However, because DOT regarded their job as fulfilled, they did not feel that there would be the interest in DOT to make them a good [lead] agency. AEC [the Atomic Energy Commission, the predecessor of the Department of Energy]. . . was eager to undertake the job, but did not have any particular agency motivation other than the aircraft and balloon capability to make measurements.[11]

In this same memo, it was stated (probably by Naugle) that

> NASA obviously has a considerable interest and motivation because of the Shuttle chlorine problem and is already working in the area. Todd [had] indicated to Stever that if there is a need for a crash effort, he felt NASA was best suited to do the job [but] NOAA is interested in being the lead agency.[12]

Todd went on to say that NOAA had done some lobbying. NOAA had given Stever a copy of a letter it sent Congress, which stated NOAA's willingness to take "responsibility to act as lead agency." Todd did not believe his agency, NSF, was appropriate to serve as the lead. Rather, he stated his view that it was NASA or NOAA that should be the lead. Naugle made it clear that NASA "was definitely interested in being the lead agency." However, NASA did not want to run a program like CIAP—one limited in scope and duration. NASA wanted "an ongoing stratospheric research program." Todd indicated he would "pass this word along to Stever."[13]

Fletcher had redistributed existing funds to set up an office. Fletcher and Deputy Administrator George Low now forcefully pursued the agency leadership position, using the office Fletcher established as what one study called a "lever." It was a way of showing that NASA was already organized to conduct an activity. To the extent there was competition between NASA and NOAA for this new program, NASA had the clear advantage in congressional support and in the freedom to maneuver as an independent agency (versus NOAA's subordinate location in the Department of Commerce).

NASA's oversight committee, the Senate Committee on Aeronautical and Space Sciences, held hearings in December 1974. As the chair, Senator Moss (Democrat, Utah) said the hearings highlighted certain facts: 1) theoretical projections that the world had a real problem, but there was little experimental evidence to back up the theory; 2) there were surprisingly few scientists working in the field of upper atmospheric chemistry, perhaps no "more than a hundred in the entire world"; and 3) "efforts to understand what happens in the upper atmosphere have been piecemeal and fragmented." Everyone agreed, Moss said, on the need for a major research program that would focus on this problem. NASA, in his view, was the right agency to lead this effort, given its "unique capabilities."[14] The decision to ban certain ozone-depleting substances would depend on what this research showed.

Congress went along with Moss's view and in June 1975 passed legislation directing NASA "to conduct a comprehensive program of research, technology and monitoring of the phenomena of the upper atmosphere."[15] This language, embodied in the fiscal year (FY) 1976 authorization bill for NASA, gave the Agency a clear mandate to perform research concerned with depletion of the ozone layer.

Stage 4—Early Implementation

Congress gave the Agency $7.5 million as a specific line-item appropriation for research in NASA's FY 1976 budget. An Upper Atmosphere Research Office (UARO) was set up within NASA's Office of Space Science and Applications (OSSA) to handle these funds. Congress also appropriated an additional $115.5 million for satellite development.

The first director of UARO was James King. He established working relationships with groups at various NASA Centers: Langley Research Center (Langley), Goddard Space Flight Center (Goddard), Ames Research Center (Ames), Johnson, and the California Institute of Technology's Jet Propulsion Laboratory (JPL). He set as UARO's short-term goal the evaluation of the potential effects of the Space Shuttle, fluorocarbons, stratospheric aircraft, and other chemical emissions on the stratosphere. Initially, NASA utilized staff members of various research projects at the various Centers. Over time, the program grew and became more directed to specific issues related to ozone depletion and ozone science.[16]

The program got under way in a highly contentious political environment. Environmentalists wished to regulate early, before the research results were in, on the basis of "the precautionary principle." The $8 billion CFC industry, led by its dominant company, Dupont, wanted to wait to see what the research said, since the economic implications were great. The official position of the Ford administration was to conduct research before promulgating regulations, and Fletcher publicly supported that view.

NASA's UARO budget grew to $11.6 million by FY 1977 and would continue to rise for the next several years. In 1978, NASA launched Nimbus 7, a satellite specially equipped with five instruments to study the upper atmosphere. Although the research and satellite efforts were still relatively new, the politics of ozone were forcing early action.[17]

BANNING CFCS IN SPRAY CANS

In 1975, a federal task force recommended banning the chemicals used in most aerosol propellants for hair sprays, shaving creams, and deodorants unless subsequent scientific evidence exonerated them. In 1977, a National Research Council report indicated that they were a subject of legitimate concern.[18] There was little connection between NASA's stratospheric R&D program and this policy. Science and regulation were subject to differing forces, timetables, and decision-making processes. However, Congress seemed to want more policy-program integration. In 1977, it passed legislation (Clean Air Act Amendments), which required NASA to issue biennial reports to Congress on the status of ozone science and what was known.[19] In 1978, Congress banned CFC use in aerosol propellants—the strongest regulation that could be accomplished at that time. Decisions on other CFC sources, particularly refrigerants, would have to wait.[20] Industry could substitute other technologies for the spray cans. Because industry said it would be more complicated to find substitutes for other uses, Congress limited the policy to particular applications in the United States.

BUILDING SCIENCE

In the late 1970s, NASA sponsored workshops and published reports on the research and status of ozone-layer depletion. It also began plans to launch a new satellite dedicated specifically to ozone-layer depletion, called Upper Atmosphere Research Satellite (UARS).[21]

The political context in which NASA implemented its program changed dramatically when Ronald Reagan replaced Jimmy Carter as President in 1981. The White House opposed regulation but supported scientific research, especially of a more basic kind. The Reagan administration saw the existing CFC policy as more than adequate. It put proposed Environmental Protection Agency (EPA) regulations on hold, and industry, which had been pondering possible substitute technologies, now relaxed a bit. Even environmentalists were relatively passive, resting on their laurels from the 1978 aerosol propellant legislative victory. The experience of NASA satellite researcher Donald Heath in 1981 characterized the new political setting. He told EPA that satellite observations showed a 1-percent loss of global ozone. EPA said Heath's views were "mildly suggestive." Heath backed off. "There were many questions, but I still believed it was real," he said. "There was so much opposition to it; I sort of let it die. I thought I'd wait a while."[22]

The Reagan administration did believe it was a governmental responsibility to perform basic research, including the study of stratospheric chemistry. The Upper Atmosphere Research Program received between $20 and $30 million a year for research (exclusive of satellite development). Moreover, NASA began planning a larger effort concerned with the global environment that made the upper atmosphere program extremely important as a first step for the Agency as a whole.

DEVELOPING AN ENVIRONMENTAL MISSION

Reagan's appointee, NASA Administrator James Beggs, came with a government and industry background. Beggs viewed the Space Shuttle as just about "operational" and NASA as being in dire need of one or more new big R&D missions. His priority was a space station, which he called "the next logical step" in space exploration. However, in 1981, he knew it would take time to "sell" such a giant program to the Reagan administration. As it turned out, he could not get the President to go ahead with the Space Station until 1984.

Meanwhile, Beggs encouraged other large initiatives. The head of OSSA, Burt Edelson, who was an engineer, was interested in the applications side of OSSA. He saw the ozone depletion research as a harbinger of missions to come and envisioned NASA with a major Earth science role. Such a mission, he believed, would require large platforms in space with multiple sensors looking comprehensively at air, land, and sea changes on Earth.

Beggs liked what he heard from Edelson and in 1982 proclaimed before the UN Conference on the Peaceful Uses of Outer Space that NASA was going to launch Project Habitat. NASA, he said, would lead "an international cooperative project to use space technology to address natural and manmade changes affecting the habitability of Earth."[23] The reaction was quite underwhelming, and in fact, negative. Other agencies and nations asked what NASA was up to—they felt blindsided and viewed NASA as engaging in bureaucratic empire-building. NASA had failed to properly brief many of the parties whose support it would have needed to get Project Habitat off the ground.

Beggs told Edelson to go back, plan a program, and build a constituency before moving forward in any visible way. Edelson appointed an Earth System Sciences Committee to explore options and included other agencies, particularly NSF and NOAA, in the deliberations. Meanwhile, outside NASA, the scientific community independently began contemplating a huge international effort, which came to be called "global change." At issue here were subjects such as desertification, ocean pollution, climate change, and others. These problems involved more than space but certainly needed the view from space.[24] The Upper Atmosphere Research Program, thus under way, was seen inside and outside NASA as critical to global change plans.

In 1984, NASA got Reagan administration approval to develop the Upper Atmosphere Research Satellite, for which NASA had been planning since 1978. UARS was seen as the principal driver in NASA's ozone depletion research and development program. It was a major technical advance over what existed and would cost hundreds of millions of dollars to develop. Since it would be some time before UARS could be deployed via a Shuttle, the NASA approach was to conduct research "along the way." The pace of R&D was governed by the development of UARS, scientific questions, the need to build a scientific cadre, and bureaucratic requirements. Science teams worked out a strategy for using UARS data when they became available, so as to accelerate the application of this knowledge as soon as possible.[25]

ROBERT WATSON AS A POLICY ENTREPRENEUR

The man in charge of NASA's Upper Atmosphere Research Program—the nonsatellite activity—was Robert Watson, a young atmospheric chemist who had been born and educated in Great Britain. After receiving his Ph.D., Watson did postdoctoral work at the University of Maryland and then the University of California at Berkeley, under Harold Johnston. He then went to work at NASA's Jet Propulsion Laboratory, managed by the California Institute of Technology (Caltech). He became involved in the Upper Atmosphere Research Program while at JPL, and in 1980 NASA moved him to Washington, DC, to manage the upper atmosphere program. He was then 32.

Watson started out as "purely academic" in his orientation, but his doctoral work focused specifically on chlorine chemistry as it related to the CFC–ozone loss issue. His postdoctoral work and JPL experience drew him increasingly to believe that stratospheric ozone loss was a real threat. He was familiar with the theories of Molina and Rowland and gave them credence. He wanted NASA to work on a problem he believed was potentially quite serious.[26]

Watson's administrative base was solid. NASA had a legislative mandate for a program and a steady budget for ozone depletion research. It controlled approximately 70 percent of the federal government's ozone-related research dollars. Watson ran a program that subsequently matured and had a growing constituency of researchers interested in the problem. Watson also had a legislative mandate to provide policy-makers with an assessment of the science every two years. This assessment aspect, legislated during the Carter administration and then carried out later by NASA, might have been seen more as a symbol than reality, but Watson saw it as an opportunity to greatly expand the usefulness of his program.

Watson developed a certain strategy that would become his research and assessment trademark. This was a "participative strategy." Because of the mandate and the relatively large resources available, NASA had the potential to play a considerable role as the lead agency, and Watson desired to do so. He wanted others aboard and used the assessment responsibility to build alliances among scientists in different agencies, and eventually nations. Watson became a proactive coordinator of scientific ozone research and assessment efforts that grew increasingly complex and far flung in scale. Jack Kaye, a NASA official concerned with ozone research, recalled Watson's skill as akin to "herding cats."[27] Watson had a rare capacity to get scientists to work together. A writer who studied the ozone-depletion issue called Watson "a master at blending the roles of bureaucrat and scientist."[28] This blending was shown most vividly in how he dealt with the assessment function.

Watson found that NASA's assessment was one of several, and that fragmentation weakened its impact on policy-makers. When he took leadership of the program, he recalled thinking that

> Before 1980, there were several assessments being done periodically in different countries. This just meant that the policymakers spent more time looking at the differences between them rather than at the similarities, even when they said basically the same thing. With one document, even if there was a range of views in it, then the international policy community had a constant base.[29]

In 1982, Watson convinced the Federal Aviation Administration (FAA) and NOAA to cosponsor the NASA assessment. He also internationalized it by enlisting the World Meteorological Organization (WMO) as a cosponsor. Subsequently, the NASA assessment was often called the NASA/WMO assessment, or even WMO/NASA assessment.[30] Watson, unlike many Washington administrators, saw "sharing credit" as a way to enhance influence.

Early on, he struck an alliance with Mostafa Tolba, head of the United Nations Environmental Program (UNEP). Tolba, like Watson, believed that ozone depletion was a serious problem and that it was important for scientists and policy-makers to come together and determine what to do.[31] UNEP was a relatively weak entity by UN standards, but Tolba seized the ozone-depletion issue as his and began sponsoring international meetings that kept the issue on the agenda of scientists and policy-makers. Watson found that UNEP meetings helped him to keep scientists focused on the issue as well.

The UNEP meetings during the early Reagan years gradually became more substantive, with an eye to negotiating an international policy agreement. The United States participated officially (via the State Department), in part because it had banned CFCs in aerosol cans and wanted other nations to follow suit.

Watson helped UNEP by sponsoring a series of 30 workshops whose findings were fed not only into the NASA assessments, but also into those international meetings. Watson involved 150 scientists from 11 countries in workshops over the 1983–84 period. He used the workshops to push for agreement on facts.[32]

In preparation for the 1986 NASA assessment, Watson expanded co-sponsorship even further, adding more agencies in the United States, more international bodies, and a West German government agency. Litfin writes that Watson's motives were more political than scientific.

The reasons for including broad representation were more political than scientific. Watson and the other scientists who saw the need for a strong international report "wanted to break down the false skepticism that wasn't based on fact, but rather on things like, 'This is only American research . . .'" . . . He attracted scientists to the workshops by emphasizing their professional value, stating that "the world's best atmospheric scientists would be there" and that "a document would come out of them that we could all be proud of." Some scientists from certain countries were invited to the workshops even if they had little to contribute, in the hope that they might stimulate interest at home. Overall, the rationale for the assessment was inherently political—to mitigate nationalistic biases.[33]

Watson felt an increasing sense of urgency, but aside from a few others (like Tolba), this feeling was largely missing from most scientific and policy discussions in the 1981–84 period. Moreover, Watson himself was conscious of the limits of scientific knowledge about ozone depletion. He sided with industry in 1984 in calling for a delay in further CFC bans pending further research and satellite observations. In early 1985, he and his superior at OSSA, Shelby Tilford, considered whether a Total Ozone Mapping Spectrometer (TOMS) unit on Nimbus satellites and high-flying aircraft (ER-2s) would be the best equipment to use or some other method might be better. Events that year made them speed up research and decide on equipment, as circumstances changed dramatically.

In January, EPA got a new leader, Lee Thomas, who was handpicked to help the agency recover from a period in which the first Reagan appointee had not only weakened EPA, but also embarrassed the Reagan administration publicly. Thomas soon decided that ozone depletion was an issue on which EPA had to take a stand. His agency relied on NASA for stratospheric research. However, it also studied the cancer risks of exposure to excessive ultraviolet rays, and Thomas viewed these risks in "black-and-white" terms.[34] That is, he believed any exposure was bad, and, therefore, EPA had to push for a maximum possible ban on CFCs.

Another change in circumstances affecting the ozone-depletion issue came in March 1985 when international negotiators adopted a "framework" convention. There was agreement among several nations, the United States included, that ozone depletion was indeed a problem and that the world should address it. This agreement, the 1985 Vienna Convention for the Protection of the Ozone Layer (known as the Vienna accord), did not have any controls and carefully recognized the sovereignty of individual states to do as they pleased. But at least there was now official international recognition of the problem.[35]

The Nimbus 7 satellite. (Source: Folder 6187, NASA Historical Reference Collection, NASA History Division, NASA Headquarters, Washington, DC.)

THE OZONE HOLE

Then came the event that jarred everyone concerned with ozone science and policy—the publication of a paper that asserted the existence of an ozone hole above Antarctica. The genesis of this paper went back to 1982. A team from the British Antarctic Survey, led by Joseph Farman, found ozone losses above 20 percent in an area of Antarctica (Halley Bay) they were studying. Existing models of ozone depletion had predicted a gradual loss—nothing as big or dramatic as reported by the British team. Trusting those earlier models, NASA scientists had questioned the reliability of satellite observations that were similarly anomalous, suspecting that they were an artifact of the instrumentation. Farman himself distrusted his data at first.

Over the next two years Farman conducted more research, as he came to believe the ozone loss was extensive. He wrote to NASA scientists at the Agency's Goddard Space Flight Center, who were responsible for the TOMS data, about his results and asked them whether they were seeing what he was seeing. He did not get a reply. Through his own research, he discovered that ozone loss varied with the season, a fact suggesting meteorology rather than chemistry might explain the loss. However, in his paper Farman specifically pointed to CFCs as the cause, while proposing that the meteorological conditions were a contributing factor. Farman also calculated that from 1957 to 1984 the surface area of the ozone layer blanketing Antarctica had shrunk by 30 percent. Farman did not sound the alarm via the media but instead went through normal scientific procedures. He finished his paper, had it peer-reviewed, and eventually published it in the May 1985 issue of

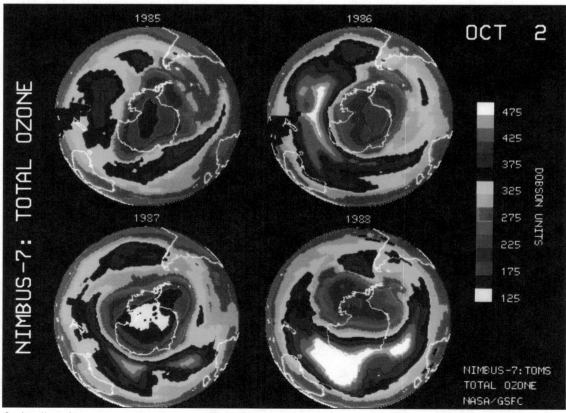

Southern Hemisphere ozone cover 1985-88, as mapped by Nimbus 7: TOMS. (NASA Image No. 89-HC-10)

Nature magazine.[36] There were NASA scientists, meanwhile, who were becoming belated believers in the "anomalous" satellite readings. They also prepared a paper supporting the view that CFCs were responsible for significant ozone depletion and planned to present the paper at a scientific conference. However, Farman's paper came out first.[37]

After a brief time, in which scientists, the media, and policy-makers digested the implications of the Farman paper, the ozone issue exploded into a huge national and international debate. Watson wanted to know why NASA scientists had not spotted the problem earlier. The scientists at Goddard told him they had assumed that all the anomalies in the TOMS data were in error, but it turned out all the anomalies occurred over Antarctica in the austral spring. Initially, NASA was chagrined that a relatively obscure British research group had achieved scientific superiority and that its satellites appeared not to have yet identified—or at least researchers using satellite data had not spotted—the hole. However, NASA then transmitted satellite images of Antarctica, and what those pictures revealed literally brought the ozone issue home to the general public and its political representatives.

The satellites showed the scale of the hole—the size of the continental United States. Moreover, as NASA official Jack Kaye remarked, "You could watch it grow, evolve like a living organism."[38] Those color images appeared on television, shown repeatedly on evening newscasts. The publicity was intense. Scientists and policy-makers wanted to know what was happening at the South Pole and whether this might be an early warning of a global phenomenon. What bothered everyone was the suddenness of the discovery, the knowledge that scientific models had assumed a gradual loss, and the staggering size of the depleted area. The possible implications for policy were clear to EPA Administrator Thomas, the environmental community, and many scientists (including Watson). There was less time to act than previously assumed. The NASA program had to become more targeted.

Fortunately, Watson had invested heavily in three scientific researchers, whose work, he believed, was of sufficient quality as to allow real progress in testing atmospheric-chemistry models. These were James Anderson at Harvard University, Crofton Farmer at JPL-Caltech, and Art Schmeltekopf at the NOAA Aeronomy Laboratory. The key now was to get them and others marching in the same direction.

Stage 5—Evaluation/Reorientation

The key scientific questions were 1) whether the hole was due to chemistry (i.e., CFCs) or nature (i.e., meteorology) and 2) whether what was happening in Antarctica was also happening globally. Watson seized the moment to try to get answers to these questions. The process began in January 1986, with the NASA assessment.

This report, *Present State of Knowledge of the Upper Atmosphere: An Assessment Report*, summarized what was known, saying little about the hole since investigations were only just unfolding. However, the report did state that existing models predicted a global ozone loss of between 5 and 8 percent by the end of the 21st century. It also predicted a loss of as much as 14 percent at the polar regions, which did have unusual meteorology. Alluding to Antarctica, the report concluded, "What was once mainly based on theoretical predictions is now being confirmed by observations." It also asked whether the ozone hole could be "an early warning of future changes in global ozone."[39]

The lead author of this report was Watson, joined by various NASA colleagues. In his preface, Watson said that 150 scientists in different countries contributed to the report. The lead reviewer was Dan Albritton, director of NOAA's Aeronomy Laboratory in Boulder, Colorado. One of Watson's most fruitful moves was to establish a personal and institutional alliance with Albritton.

When the leadership of Upper Atmosphere Research Program was being considered by Congress, NOAA had put in a claim to be the lead agency but had lost to NASA. Watson might have ignored NOAA, a potential rival. Instead, he made it a partner, and the Watson-Albritton alliance proved important not only in responding to the ozone hole crisis, but also in communicating its threat to policy-makers. Albritton, like Watson, had an unusual skill. According to Kaye, Albritton was exceptionally competent in communicating technical information to policy-makers. It was not just his verbal capacity, however, but his ability to use hand-drawn viewgraphs that helped to demonstrate his points in a clear, understandable, and easily remembered way.[40]

Albritton did not have much discretionary money to contribute to what became a crash effort in ozone research and assessment, but he had his own abilities and those of the talented people in his lab. In 1986, Watson moved from a research manager to a research mobilizer and science advisor to policy-makers. Albritton was

Robert Watson of NASA examines TOMS ozone data. (NASA Image No. 87-HC-223)

his chief partner in all these roles. NASA gave Watson remarkable autonomy and enough resources. NASA's overall planning made the study of ozone depletion the vanguard of what it came to call publicly its "Mission to Planet Earth (MTPE)."[41] MTPE emerged gradually, becoming a formal office in 1988.

THE ANTARCTIC EXPEDITION

In March, Watson, Albritton, and others met in Boulder and decided to conduct a field expedition to Antarctica as soon as possible. With Albritton's help, Watson hurriedly dispatched a 13-member team of scientists to Antarctica during the August–September period when the depletion seemed to be most pronounced. He placed Susan Solomon, a NOAA scientist from Albritton's lab, in charge. Known as NOZE (National Ozone Expedition), the expedition's purpose was to explain the ozone hole and determine which of the various theories accounted for it. Funding came mainly from NASA, but NOAA, NSF, and the Chemical Manufacturers Association also contributed. The team took balloon and ground-based measurements and also had the benefit of satellite data.[42]

At the end of NOZE, Solomon held a press conference from Antarctica. Although there still was much work to do analyzing the data, the NOZE team felt that public alarm about the ozone hole required them to say something (rather than waiting, as had Farman, until their work had cleared the lengthy peer-review, publication process). Solomon declared, "We suspect a chemical process is fundamentally responsible for the formation of the hole."[43]

There were caveats, and by no means did Solomon say the data were conclusive. But Solomon's statement about causation received a negative reaction from many scientists who favored a meteorological explanation and from industry, which thought the statement much too premature. In November 1986, critics of NOZE aired their views in a special edition of *Geophysical Research Letters*. Many of those who published opinions in the journal wanted a higher standard of proof than existed at the time.[44]

Watson decided that there would have to be a second expedition to settle the scientific questions regarding the cause of ozone depletion. Meanwhile, the national and international policy debate resumed with an urgency that matched the fervent scientific search for an explanation. Indeed, the regulatory policy process, galvanized by the ozone–hole issue, was now moving faster than that of the science.

POLICY DEBATE

In December 1986, international negotiations on a possible ozone treaty resumed in Geneva, Switzerland. There had been a 17-month hiatus since the 1985 Vienna accord. In the wake of the ozone-hole discovery and the publicity it engendered, more nations now stood ready to negotiate a treaty with some controls.

Within the Reagan Administration there was considerable debate. One side wanted no regulation until the scientific uncertainties were more fully resolved—a view Reagan's science adviser, William Graham, espoused. The other side, championed by EPA Administrator Thomas, argued for maximum precautionary controls with a ban on 95 percent of the CFCs that were being produced, thus forcing industry to come up with substitutes sooner than later.[45]

In May 1987, Interior Secretary Donald Hodel caused an uproar when he was quoted as saying that if ozone depletion made ultraviolet radiation increase, people could simply wear sunglasses and hats and use suntan lotion. "People who don't stand out in the sun—it doesn't affect them," he declared. Hodel's remarks brought widespread criticism, ridicule, and scorn. The media had a field day. Environmentalists appeared at press conferences wearing sunscreen, hats, and sunglasses. Congress held hearings, putting more pressure on the administration to "do something." Secretary of State George Shultz distanced himself from the antiregulatory opposition and said the international negotiations would continue. President Reagan became involved and gave his assent to finding a resolution for the issue. The administration worked out a policy compromise that asked for a 50-percent reduction in CFC production by 2000.[46] The United States sought to make this position a global policy.

A SECOND ANTARCTIC EXPEDITION

Watson and Albritton were asked to serve as principal science advisers to the U.S. delegation to the international negotiations. Watson saw two challenges. First, he had to organize another antarctic expedition to settle the chemistry-versus-meteorology argument. Second, he had to deal with an internal problem at NASA. Donald Heath, from the NASA satellite program, had testified before Congress that his evidence showed an ozone loss of 4 percent over the entire planet within a seven–year period. Watson had testified that the evidence did not support such a statement.[47] Watson believed he was correct and that the scientific community's credibility depended on stating not only what scientists knew, but also what they did not know. Moreover, he also adhered to the view that scientists had to present a united front when advising policy-makers if they wished to have influence. He wondered how he could build a consensus among different countries, when he could not effect a unified voice at NASA. Hence, in addition to the antarctic expedition, he sponsored a large, international scientific panel, which he called the Ozone Trends Panel, to assess existing data on the global impact of ozone depletion. The panel's mission was to dig into the data in some detail and look at the resulting trends to determine whether they were real or might be due to uncertainties, inaccuracies, or a degradation of the observing systems—an important point if the aim was to reveal a subtle trend within a data set that was highly variable.[48]

First, there was Antarctica. With Albritton's help, Watson organized what was called the Airborne Antarctic Ozone Experiment (AAOE). Watson wanted results and to get them he decided to "throw everything" into this project.[49] Not only would there be a larger number of scientists deployed, but the research team would have more and better equipment. This antarctic expedition cost $10 million, with NASA paying the most, and with cosponsorship from NOAA, NSF, and the Chemical Manufacturers Association (industry's representative on the research team). A total of 150 scientists and associated support personnel were involved, representing 19 different organizations and four countries. Among others, the United Kingdom Meteorological Office played an important role providing for its own expenses.[50]

Watson believed it was critical to get measurements in the stratosphere, where ozone had begun to disappear. The first expedition obtained satellite and ground data. The data were suggestive but inconclusive. He needed an airplane that could fly 13 miles above Earth. There was only one type of plane at NASA that could fly at that altitude, a converted spy plane called the ER-2. NOAA's Schmeltekopf had suggested the need for such a plane as early as the March 1986 meeting in

Boulder that proposed NOZE. Acting upon this suggestion was possible because Watson and Tilford, Watson's superior at NASA, had directed funds for research using in situ instruments on the ER-2 to investigate ice cloud formation at temperatures very similar to those in the antarctic vortex. This work demonstrated the potential value of the ER-2 to the study of ozone depletion. Getting the airplane and using it in Antarctica was not easy, however.[51] In order to acquire the plane, Watson had to bargain with another division of NASA. Meeting some resistance, he declared, "Look, this isn't just a science problem. It's a problem that has important policy implications that a lot of people care about."[52]

Watson got his plane. In gearing up rapidly for the expedition, he was helped by his previous NASA experience. NASA's larger organizational needs at this time also aided his efforts. In 1986, NASA had suffered a blow from the *Challenger* disaster that took the lives of seven astronauts, including the first teacher in space. In 1987, James Fletcher returned to NASA as administrator to help the Agency recover. Fletcher had been in charge when the Upper Atmosphere Research Program had begun. Sally Ride, the first U.S. woman in space, did a study for Fletcher to identify post-*Challenger* NASA priorities, and the Mission to Planet Earth got equal billing in her report with the missions from Earth.[53] Counting satellite expenditures during this period, Watson estimated total NASA spending on ozone R&D at $100 million in 1987.[54]

Once he had his ER-2, Watson needed to have instruments installed. He contacted James Anderson, a Harvard University atmospheric scientist, to whom he had directed funding for some time and who had built a balloon-borne instrument. He asked Anderson to adapt the instrument for the ER-2. Watson also requested that he develop an instrument to measure ozone-depleting chemical reactions 13 miles above Earth. Anderson said he needed $400,000 for the task, now seen as urgent.[55] Watson provided the money.

In August 1987, the Airborne Antarctic Ozone Experiment began. The project manager was Estelle Condon of NASA's Ames Center, the site from which the ER-2 was obtained. As in the case of NOZE, Watson selected a NOAA employee as the principal scientist—A.F. Tuck of the NOAA Aeronomy Lab. Along with the ER-2, the expedition employed balloons, satellites, a DC-8 flying laboratory, and other equipment.[56] The expedition was a major logistical feat, one that was both helped and hindered by the NASA culture. NASA's experience with large missions, often under tight deadlines, operating in a harsh environment was advantageous.[57] In contrast, the resistance by some NASA officials to Watson's granting so much authority to a NOAA scientist was a hindrance. Also, Watson took risks by hand-picking his team, funding certain research without peer review, and using funds that he had saved for an emergency.[58]

As the project proceeded, Watson and Albritton spent much of the time on airplanes. They worked together and separately, dividing labor as necessary. They were research managers and also science advisers to the international negotiations. These negotiations were now moving on a schedule that would not wait for the antarctic expedition or Ozone Trends Panel to reach final judgments.

THE MONTREAL PROTOCOL

The ozone hole was very much on the minds of negotiators.[59] The delegates saw the hole over Antarctica as a warning, even though conclusive scientific findings were unavailable. Watson and Albritton told the delegates what they knew, and what they did not know. However, there were many nonscience issues that still had to be addressed.

The United States, led by the State Department and EPA, had considerable leverage in the international meetings that resulted ultimately in the Montreal Protocol on Substances That Deplete the Ozone Layer. The United States was the dominant nation in CFC production. It wanted a global solution for a global problem. Advocates of the Protocol perceived industry as being capable of coming up with substitutes. Developing nations would need technical assistance in converting to CFC substitutes and would require financial help.[60]

In September 1987, Watson and Albritton flew up to Montreal during the final negotiations to present the participants with the preliminary results of the expedition. However, before the final scientific facts were in from Antarctica, the Montreal Protocol had concluded. It called for a 50-percent cut in worldwide CFC production by 2000. A total of 43 nations initially signed the accord. Aware that more scientific information would soon be available, the delegates agreed to amend the protocol if the new scientific evidence warranted changes.[61]

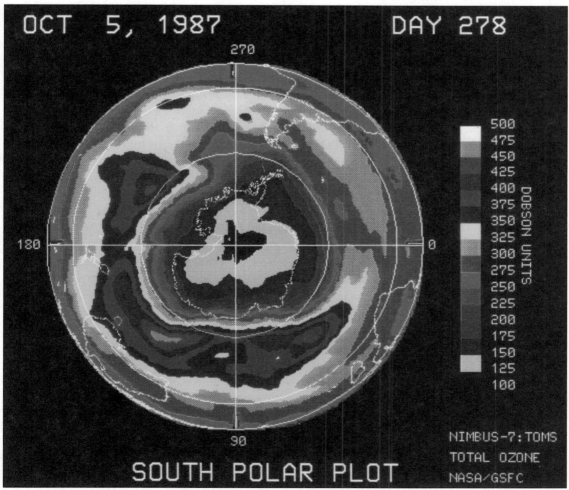

Southern Hemisphere ozone cover in 1987, as mapped by Nimbus 7: TOMS. (NASA Image No. 89-HC-574)

CONCLUSIVE RESULTS

In October, leaders of the second antarctic expedition studied their results. The evidence was now much stronger, because it showed that ozone decreased in the stratosphere as CFCs increased. The observations backed the theory. They announced that almost half the ozone over Antarctica had disappeared in August and September and that CFCs, rather than natural weather conditions, were strongly implicated as the cause. The weather at the South Pole merely exacerbated the problem. This second expedition also provided the "smoking gun" for which participating scientists had been looking—a very clear "anti-correlation" between chlorine monoxide (the chemically active form of chlorine in the stratosphere) and ozone. That is, the more chlorine, the less ozone! Critical to this finding were two instruments on the ER-2—the chlorine monoxide instrument from Anderson's group at Harvard, and an ozone instrument from the NOAA Aeronomy Lab.[62]

The following month, the Ozone Trends Panel met—100 scientists from different nations. Watson had wanted the panel to settle his dispute with NASA colleague Donald Heath, but the participants instead decided to postpone discussing that issue and review the antarctic findings. The Ozone Trends Panel backed the second expedition's results, agreeing that the data showed that CFCs were most likely at fault for ozone loss.[63] Science thus legitimatized the policy decision of Montreal.

In February 1988, the AAOE science team convened in Estes Park, Colorado. Upon reviewing the second antarctic expedition's findings, the team realized that there were similar conditions in the Arctic that could contribute to ozone loss there as well.[64] Meanwhile, in March, the U.S. Senate ratified the Montreal Protocol. That same month, Watson reconvened the Ozone Trends Panel to deal with the global impacts of ozone depletion question. The panel concluded that the satellite instruments on which Heath had relied for his evidence had suffered degradation, and thus his findings were suspect. However, ground-based stations in the Northern Hemisphere did indicate Heath was on the right track in certain regional measurements. There were indications of a 1.7- to 3-percent loss in the Northern Hemisphere, perhaps as far south as New Hampshire.[65]

These findings, along with the Estes Park discussions in February, startled many observers, because the Northern Hemisphere was much more densely populated than Antarctica. Dupont, the world's largest CFC manufacturer, announced it would cease production as soon as substitutes became available. On 5 April, President Reagan signed the Montreal Protocol, saying it created incentives for new technology. In August, EPA ordered domestic CFC reductions in line with the international agreement.[66]

Thus, the Montreal Protocol served as an end to one decision-making process involving science and policy and the beginning of another. New scientific findings triggered further policy action, and the momentum for change continued. The Montreal Protocol included measures for subsequent review and amendment based on further scientific research. The policy-makers understood that they were making policy and setting rules to contain CFCs under conditions of uncertainty. The delegates virtually invited the scientific community to reduce uncertainties in order to guide policy.

The existing agreement, ratified by the U.S. Senate in 1988, called for a 50-percent reduction in CFCs by 2000. Should that standard be tightened to 85 percent or 95 percent? Policy-makers wanted to know the final standard—and so did industry, which was hard at work on substitutes and feeling not only the pressure of regulation, but also competition within the private sector as to who could come up with substitutes first. There was a definite sense of political momentum in the wake of the Montreal Protocol. Science was responding to policy pressures and also creating pressures of its own.

LOOKING NORTH

In October 1988, NASA began serious planning for an arctic expedition. The intent, once again, was to employ an array of equipment—satellites, high-flying airplanes, balloons, and ground stations. As before, NASA would take the lead but would work closely with NOAA, as well as with NSF, industry, and others. Of particular interest were polar stratospheric clouds (PSCs), which scientists hypothesized were contributing to ozone depletion. Richard Stolarski, a NASA scientist at Goddard, said, "Our work in the Arctic . . . will focus on the polar stratospheric cloud, if and when it develops. We want to know what chemicals go into the cloud, how they are processed inside the cloud, and what comes out."[67]

Stage 6—Amplification

As plans to expand the ozone-research activity were made, they moved into the context of NASA's ambition for even greater participation in environmental studies. For example, from the time the ozone program was young, NASA had conceived of developing a satellite with instruments more advanced than TOMS-Nimbus or the Upper Atmosphere Research Satellite. This device was in the works by the late 1980s, just as ozone and other environmental issues were becoming top priorities for the nation and NASA.

NASA scientists had been thinking of creating a comprehensive satellite system for some time, but the ozone-research work made them think bigger, sooner, and in ways that linked various environmental issues. Also, in the wake of the *Challenger* disaster, NASA had found its role in ozone research and policy therapeutic. In her 1987 report for the Administrator, Sally Ride spoke of NASA's mission on the home planet as being equally important to those missions of planetary science and even human spaceflight. NASA scientists planned to launch an Earth Observing System, a series of pairs of huge platforms with multiple space sensors that could look at land, air, and sea simultaneously (and their interactions). In the years ahead EOS would constitute the heart of the Agency's Mission to Planet Earth. It would include ozone measurements, but it also would examine climate change and other environmental problems. Buoyed by the excitement and interest generated by its work in ozone research, NASA decided to accelerate ongoing plans for EOS. It asked Congress for $20 million in its FY 1989 budget to enhance R&D, with plans to launch EOS perhaps as early as 1995.

The year 1988 turned out to be the point at which climate change moved onto the White House agenda. The summer was extremely hot and NASA scientist James Hansen testified about climate change threats before the U.S. Senate. He gained considerable attention when he said that he was very confident that the "signal" of climate change had been detected and, furthermore, that human activities were almost certainly the major cause.[68] Hansen spoke for himself, not NASA, but NASA received both credit and blame for publicizing the threat. Drought in the summer of 1988 caused considerable economic hardship in many parts of the country. The UN established an Intergovernmental Panel on Climate Change (IPCC), modeled on the Ozone Trends Panel, as a way to prompt scientists from different countries to reach consensus, sooner rather than later, on global climate threats.

Nineteen eighty-eight was also the year in which President Reagan's science advisor named a new interagency committee called the Committee on Earth Sciences (CES). With NSF, NOAA, and NASA leading the way, CES produced a report calling for the set up of the U.S. Global Change Research Program (USGCRP).

Global environmental issues became part of the 1988 presidential campaign. Vice President Bush captured the issue from his democratic rival, Michael Dukakis. In speeches, Bush said he would be the "environmental president" and would use the "White House effect" to counter the "Greenhouse effect." Never before had global environmental issues received so much attention in a U.S. presidential campaign. Ozone depletion was part of the rhetoric and call for action.

EXPANDING GLOBAL ENVIRONMENTAL RESEARCH

In January 1989, George H.W. Bush became president. Soon after, he appointed Richard Truly NASA Administrator. He also appointed D. Allan Bromley to be his science advisor. In a meeting with Bromley, Bush told him to select a few particularly significant areas of science that he could back with presidential leverage. Back in 1988, NASA, NSF, and NOAA had formulated their interagency endeavor, the U.S. Global Change Research Program (USGCRP). This interagency effort incorporated the research of Mission to Planet Earth, now an office, and EOS, the key project of MTPE. A report on USGCRP was already sitting on Bromley's desk awaiting his attention when he became the White House science advisor. Bromley decided to make global environmental change Bush's first presidential priority in science and technology. The significance of this presidential initiative to NASA's interest became increasingly apparent.

In July 1989, in marking the 20th anniversary of the Moon landing, the President publicly endorsed NASA's Mission to Planet Earth. However, Bush made a return to the Moon and traveling on to Mars his first space priority. Senator Al Gore, then the chairman of Senate Subcommittee on Science, Technology, and Space, said that the Mission to Planet Earth was America's true priority in space. All this discussion provided a context in which NASA's ozone-research endeavor, the leading edge of NASA's expanded Mission to Planet Earth, operated. The Upper Atmosphere Research Program was subsumed under MTPE. Subsequent events provided an extra imperative for NASA to move ahead with its ozone-research program.

ARCTIC EXPEDITION

In January and February of 1989, NASA, along with its allies, began its arctic research with the Airborne Arctic Stratospheric Expedition (AASE). The expedition reported that the region had the same kind of disturbed atmospheric chemistry that already had destroyed part of the ozone layer over Antarctica. Watson said "the incredible perturbation" found in the chemistry of the arctic stratosphere "is a strong message to the policy makers." The mission lasted six weeks. Led by NASA, the project's participants included NOAA, NSF, the Chemical Manufacturers Association, and others. NASA's Estelle Condon, the project manager mainly responsible for making most of the arrangements allowing NASA airplanes and scientists to work in the field, called the expedition a "resounding success." Although the expedition found a chemical threat to ozone, it did not find clear evidence that ozone-loss had already occurred over the north polar region, and the scientists could not predict with certainty that a substantial loss of ozone would occur.[69] Tracking ozone depletion in the Arctic was difficult because the arctic stratosphere was warmer than the stratosphere above Antarctica; hence, fewer polar stratospheric clouds formed to accelerate the depletion reaction. Further, the arctic polar vortex, an air mass in the upper atmosphere held together by cold polar winds, was less stable, preventing the formation of large depleted regions. It was also, thus, less predictable than the antarctic polar vortex. The scientists knew that there were differences in atmospheric conditions at the two poles and hoped to continue research to learn more about the ozone dynamics at the North Pole.[70]

POLICY CONCERN

As the arctic scientific expedition ended, the 12 countries of the European Community decided to go beyond the existing international agreements and phase out the production of ozone-depleting chemicals by the end of the century. President Bush quickly endorsed the idea of phasing out the chemicals by the century's end if alternatives could be found. In early March 1989, Britain hosted an international conference involving 123 nations. Although some countries wanted to wait, 20 additional countries said they would join the Montreal Protocol.[71]

ACCELERATING OZONE SATELLITE DEVELOPMENT

As worldwide concerns about ozone depletion grew, new data from an old spacecraft spurred policy action. This policy activity, in turn, pointed out the need for additional research. In October, NASA reported that its TOMS instrument on the Nimbus-7 satellite had confirmed that the ozone hole over Antarctica observed in 1989 equaled the record-setting hole observed two years before, in 1987.

Given the importance of the Nimbus-TOMS data, NASA was worried about the age of the satellite—11 years. NASA decided to arrange for a succession of polar-orbiting ozone mappers to fly on board new spacecraft, starting with one to be launched by the Soviets in 1991. The instrument, to be flown on a Soviet Meteor 3 satellite, would provide data at least until mid-1993.

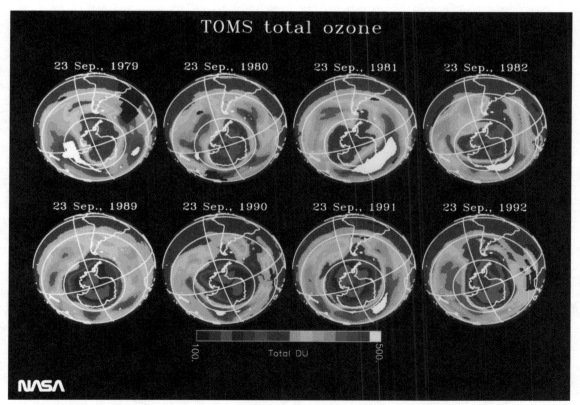

Nimbus 7 TOMS data of Southern Hemisphere ozone cover, 1979-92. (Source: Folder 6187, NASA Historical Reference Collection, NASA History Division, NASA Headquarters, Washington, DC.)

The next device would be launched as a satellite payload on board a U.S. Scout-class booster. In addition, the Japanese accepted a TOMS for their Advanced Earth Observing Satellite (ADEOS) that they planned to launch in 1995.[72]

1990—CONTINUED MTPE EXPANSION

In 1990, Mission to Planet Earth, with a presidential imprimatur behind it, took off. Funding went up 5.1 percent for Earth observations, from $610 million to $642 million. Part of this money was allocated to NASA's new satellite dedicated to ozone research, the Upper Atmosphere Research Satellite, which had been in development for some time and was nearing readiness for launch. In addition, NASA stepped up its work on the long-term centerpiece of the Mission to Planet Earth office, the EOS. It had become clear that while EOS would include ozone research, its prime focus would be climate change.

The goals of EOS were popular in Congress, but the projected cost—$17 billion over the 1990s—shocked some legislators. It remained to be seen whether NASA could secure that much money. However, the fact that NASA was thinking in such large terms reflected how far the Earth observation effort had come. EOS was a huge expansion of the ozone-research work.

Meanwhile, scientists confirmed that significant ozone depletion had occurred over the North Pole during the winter of 1989. "We're starting to see all the things going on in the northern hemisphere that go on in the southern hemisphere," said Mark Schoeberl of NASA's Goddard Space Flight Center.[73]

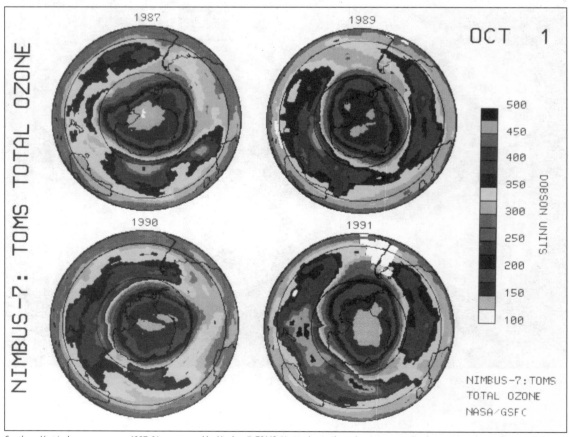

Southern Hemisphere ozone cover 1987-91, as mapped by Nimbus 7: TOMS. Notice the similarity between ozone levels in 1987 and 1991. (Source: Folder 9794, NASA Historical Reference Collection, NASA History Division, NASA Headquarters, Washington, DC.)

On 25 July, NASA signed an agreement with Russia to fly NASA's TOMS on the Russian Meteor-3 spacecraft scheduled for launch in 1991. The presence of a TOMS instrument on the Meteor-3 flight would enable it to gather critical environmental data about the yearly variability of the ozone hole over Antarctica. Meanwhile, the existing TOMS Nimbus-7 instrument revealed in October 1990 that a hole again had opened over Antarctica, and the depletion of ozone there seemed to be nearly as severe as it had been in 1987 and 1989 (the worst years).[74]

UARS: VANGUARD OF THE MISSION TO PLANET EARTH

On 12 September 1991 the Space Shuttle *Discovery* launched the $750-million Upper Atmosphere Research Satellite (UARS). NASA hailed it as the first new element in its Mission to Planet Earth. With 10 different instruments, the 7-ton spacecraft was the largest, most complex and expensive environmental satellite ever built. It was designed to provide detailed information on the chemical processes taking place in the upper atmosphere, including details on how man affected the atmosphere. Its primary aim was to gather extensive data on the planet's threatened ozone layer. "Data from the spacecraft should provide the ammunition people will need to make more informed environmental policy decisions," commented Carl Reber, a Goddard project scientist for the UARS mission.

The UARS mission development had begun prior to the formal establishment of MTPE, but NASA managers incorporated it into MTPE when President Bush authorized MTPE. Shelby Tilford, director of the Earth Sciences Division of OSSA at NASA Headquarters, called UARS a precursor to EOS.[75] Len Fisk, associate administrator for OSSA, also heralded the broader significance of UARS. He declared, "A turning point in global history has just occurred. We can now affect the global environment and that is something that's not going to change. We will need to monitor and understand this for years to come." In addition to direct measurements of the atmosphere, the UARS findings would be used by 10 international teams to improve theoretical models and predict changes in the upper atmosphere.[76]

Artist's rendition of the Upper Atmosphere Research Satellite (UARS). (Source: NASA Fact Sheet HQL-207, Folder 8653, NASA Historical Reference Collection, NASA History Division, NASA Headquarters, Washington, DC.)

NASA timed the launch of UARS to coincide with the breakup of the ozone hole in 1991 and observe a full cycle in the following year. UARS provided a novel and intense view of what was taking place. "The satellite meant that at the same time we get three-dimensional pictures of the ozone hole, we'll get three-dimensional pictures of the [chemicals] that are forming it," said Joe

McNeal, a UARS program scientist. "Having the winds coupled with the chemistry is extremely revolutionary," stated Tilford.[77]

On 8 October 1991 NASA announced that in 1991 severe ozone depletion had developed over Antarctica for the third consecutive year. The next day, NASA reported that ozone levels in the antarctic region had fallen to the lowest values ever observed during its 13 years of satellite monitoring. Additional reports surfaced two weeks later, based on data analyzed by the Ozone Trends Panel. The panel indicated a thinning of the ozone layer over the United States had occurred for the first time.

Such findings galvanized international diplomacy, industry's choice to develop CFC substitutes, and NASA's decision to launch another Northern Hemisphere expedition.

NASA SOUNDS AN ALARM

In October 1991, the second Airborne Arctic Stratospheric Expedition got under way. Earlier NASA, NOAA, and others had debated whether the Arctic or Antarctica would have priority. The prevailing view was that an arctic expedition made more sense, as there was greater variability in the arctic region and thus a greater need for detailed characterization.[78] Organized and largely funded by NASA, the expedition involved 100 scientists from NASA, NOAA, NSF-National Center for Atmospheric Research (NCAR), many universities, and an industry group. The project manager of the arctic expedition was NASA's Michael Kurylo, who assumed command of NASA's atmospheric research program as Watson moved to a higher position within NASA. The chief scientist for the project was Harvard chemist James Anderson. The team used the high-flying ER-2, the commercial-airliner-like DC-8 aircraft, as well as satellites and instrument–equipped balloons.

The initial flights originated in Fairbanks, Alaska, in October. However, because of ice on the runways, the ER-2 could not operate in Alaska during the winter. In early 1992, the flights began going out from Bangor, Maine. On 11 January satellite data indicated high levels of chlorine in the northern latitudes, including regions above such cities as London, Moscow, and Amsterdam. On 20 January, a converted ER-2 took off from Bangor and made an unprecedented flight into the center of a polar vortex. The ER-2 made in situ measurements with instruments that could sample the air outside the aircraft in real time, and it also gathered air samples that could be analyzed on the ground later.[79]

When the scientists looked at the data measured on the aircraft, they were surprised. The air samples showed the highest levels of chlorine monoxide ever measured in the arctic region. NASA's most sophisticated and latest ozone-measuring tool, UARS, seemed to confirm the findings: chlorine monoxide molecules, thought to be derived primarily from CFCs, were building up in the vortex.[80]

This information verified what the scientists had hypothesized based on data collected by the 1989 arctic expedition—that active chlorine was being liberated from the less active forms of chlorine formed from the breakdown of CFCs in the stratosphere, and just as in the antarctic region the amount of this active chlorine was increasing inside the arctic polar vortex. Also, as was the case with Antarctica, this build-up could well destroy other parts of the atmosphere's ozone layer, with a potentially negative impact on human health and agriculture.

At the beginning of February 1992, expedition leaders debated whether to wait two months until the expedition's end, to make sure their fears were valid. But by then—if those concerns were correct—an ozone hole over the Arctic would have caused damage to an unsuspecting public. There was risk in speaking up and risk in saying nothing. Some of the researchers urged caution. There was still uncertainty. Others wanted to act. In 1989, after the Airborne Antarctic Ozone Experiment, the public-testimony rule by which the participants agreed to abide was, "If we are arguing about it, it doesn't belong in a press release or congressional testimony." In 1992, that rule was not applied, and NASA decided to issue a warning.[81]

On 3 February, the leaders of the expedition, Kurylo and Anderson, appeared at a news conference at NASA Headquarters in Washington, DC. They conveyed the message that an ozone hole over the Northern Hemisphere "was increasingly likely" and had to be taken seriously. "We're not concerned with just remote regions now," said Kurylo. "What we're dealing with extends to very populated regions." With ozone loss increasing, a 30-percent loss by March was possible. "Everyone should be alarmed about this," warned Kurylo. Anderson went beyond science to public policy by saying, "We must work in a concerted way to speed up controls."[82]

MEDIA AND ENVIRONMENTALISTS REACT

The reaction from environmentalists was immediate and strong: "It's frightening," said Liz Cook of Friends of the Earth. "If the phenomenon ever occurs on a broader scale, it could be the final curtain call for life on the planet," said Karen Lohr of Greenpeace.[83] The media were no less alarmed, and influential media called on President Bush to accelerate the phaseout of ozone-depleting chemicals. A *New York Times* editorial entitled "The Ozone Hole Over Mr. Bush's Head" termed ozone depletion an issue of global importance and noted that, "[t]he life-protecting ozone layer may now be thinning above President Bush's summer home in Kennebunkport, Maine." In an editorial called "The Vanishing Ozone Layer," the *Washington Post* declared, "Once again, it turns out that the protective ozone layer in the sky is being destroyed faster than even the pessimists had expected." *Time* magazine's cover headline was "Vanishing Ozone: The Danger Moves Closer to Home."[84]

REACTION BY PRESIDENT GEORGE H. W. BUSH

Spurred by the NASA news conference, Bush's science adviser, Allan Bromley, met with various officials in the Bush administration to discuss what action, if any, was required. Bill Reilly, EPA's administrator, was active in promoting an accelerated phaseout. John Sununu, Bush's former chief of staff and an opponent of such a move, was no longer in government. Sununu's absence aided Reilly. Bush signaled a willingness to consider a faster timetable. Reilly then told the media that the president would act to speed the phaseout of ozone-depleting CFCs by three or four years. The *Washington Post* predicted Bush would announce his move at an April meeting where the signatories of the Montreal Protocol would convene. An industry spokesman said that the leading firms could find substitutes faster if necessary.[85]

SENATOR GORE PUSHES FOR LEGISLATION

Senator Al Gore, chairman of NASA's oversight committee, an environmentalist and an aspiring presidential candidate, saw a political opportunity. In November 1991 his legislation promoting an accelerated phaseout timetable for the Bush White House, operating through Republican lawmakers, was defeated. Now, he said, Bush had a "wake-up call" thanks to the "ozone hole. . . pointed to and predicted above Kennebunkport." It was about time for President Bush "to think seriously about doing something," Gore charged.[86]

Gore took the floor of the U. S. Senate to introduce a bill to halt CFC production by 1995. He termed the information in the NASA news conference "an immediate, acute, emergency threat." Following the debate, the U.S. Senate called for a halt as soon as possible, not specifying a date, but voting 96-0 in favor of speed-up.

PRESIDENTIAL ACTION

On 11 February 1992, Bush announced that he was ordering American manufacturers to end, by 31 December 1995, virtually all production of chemicals that destroyed ozone.[87] Under a provision of the Clean Air Act, Bush had the power to direct a change from the previously established year 2000 Montreal Protocol deadline, when circumstances merited such a move. Gore's response was, "better late than never." He again referred to the "ozone hole over Kennebunkport" as the reason for Bush's change of heart.

MEDIA DRUMBEAT

The President's decision and Gore's continuing volley fanned the flames of media attention. The ozone hole was now big news, a crisis, and one announcement after another of dire consequences was made. On 8 February, the *Washington Post* reported that a new UN study had linked increased UV rays from the Sun to researchers' "projections" of "300,000 new cases of skin cancer per year by the turn of the century," as well as "an increase of infectious diseases, including AIDS."[88]

On 17 February 1992 when *Time* magazine capped the media barrage with a cover headline entitled "Vanishing Ozone: The Danger Moves Closer to Home," its lead article pointed to "overwhelming" evidence that the stratospheric ozone layer "is being eaten away by manmade chemicals far faster than any scientist had predicted." The situation was dire; *Time* warned, "This unprecedented assault on the planet's life support system could have more horrendous long-term effects on human health, animal life, the plants that support the food chain, and just about every other strand that makes up the delicate web of nature."[89]

AN EMBARRASSED NASA

But in early March, the dreaded ozone hole over the Northern Hemisphere failed to materialize as predicted. Data from UARS showed that the concentrations of ozone-destroying chlorine monoxide within and around the polar vortex had declined significantly since the peak in January, according to NASA's JPL scientist Joe Waters. In January, UARS detected concentrations of chlorine monoxide at 2 parts per billion (ppb) within the atmospheric vortex that swirled around the North Pole. In February, when the satellite's orbit allowed it to look at the vortex again, chlorine monoxide levels had dropped to below 1 ppb and continued to fall.

Arlin Krueger, the NASA scientist in charge of tracking data from the Total Ozone Mapping Spectrometer, acknowledged that TOMS had found absolutely no indication of an ozone hole opening over the Northern Hemisphere.[90] He declared, "I can tell you categorically there is no ozone hole over Kennebunkport. There never has been an ozone hole over Kennebunkport, and I don't really expect one."[91] NASA had never forecast an ozone hole over Kennebunkport—the Kennebunkport reference came from Gore and others. However, due to the intense media coverage, many people blamed NASA for being the source of the reference.

COMPLAINTS FROM NOAA SCIENTISTS

The flap over the arctic ozone hole brought to the surface tensions in the organizational alliance that NASA had carefully constructed. NASA was the de facto lead agency, and others did not necessarily appreciate how they were being led, especially those at NOAA. Melvyn Shapiro, a meteorological research scientist in NOAA's Environmental Technology Laboratory in Boulder, took the occasion of a media visit to express his opinions. He harshly criticized those who downplayed natural ozone variations in favor of the CFC explanation. Implying that the arctic ozone affair was a case of "Chicken Little research," he castigated those who exploited "a doomsday scenario" to "get a lot of money." He charged, "Research organizations are in great competition with each other to get the politicians' ears and obtain the necessary resources." He did not mention NASA by name, but it was clear to which organization he was referring.

David Hofmann, senior scientist in the Ozone and Aerosols Group of NOAA's Climate Monitoring and Diagnostics Laboratory, did name NASA and complained that the Agency had given too much attention at its news conference to a CFC explanation of arctic ozone depletion. "I couldn't understand why NASA didn't come out and say that this could be a very unusual year because of the volcanic eruptions," he said, further commenting that "maybe what we're seeing is something we'll never see again. Instead, they [NASA] seemed to imply, you know, that this is the start of something really big. That really wasn't very wise. If there's major ozone depletion seen this year, it's quite likely that it is related to the volcano."[92]

Shapiro was more blunt, complaining that "this [ozone issue] is about money. If there were no dollars attached to this game, you'd see it played in a very different way. It would be played on intellect and integrity. When you say the ozone threat is a scam, you're not only attacking people's scientific integrity, you're going after their pocketbook as well. It's money, purely money."

NASA had gotten a good deal of favorable media attention for its visible antarctic ozone role. Now it took the heat for its arctic experience. Writer Micah Morrison noted strains in NASA's relationship with NOAA and NSF-NCAR because of the arctic false alarm. He noted that NOAA and NCAR were "the junior partners" in the program. "NASA is the 800-pound gorilla in the ring," said another scientist involved in the expedition, who insisted on anonymity. "You either go along with the gorilla or you stay out of its way."

"According to several scientists," said Morrison, "there was opposition within the arctic research group to releasing the preliminary findings in early February." Some project members were not even told of the upcoming NASA news conference; others were instructed to keep quiet to reporters. Dissenters felt "muzzled," Morrison reported, by the project's leadership.

While there was certainly the appearance of a NASA-NOAA split, the reality was more complicated. The Watson-Albritton alliance remained intact as to the objectives. In fact, NOAA scientists involved in the NASA-NOAA relationship felt their integrity had been impugned and demanded (and received) an apology from Shapiro. What the media did not make clear was that NOAA was chiefly represented by its Aeronomy Lab in its dealing with NASA on the ozone issue and that the critics within NOAA came from other labs. Also, there were professional disputes. Meteorologists in NOAA and NASA felt their dissenting views had been ignored by the dominant atmospheric chemists at the time the arctic ozone warning was given.[93]

When the arctic ozone-warning issue exploded in the media and became a factor in presidential politics, there were those on the arctic team who urged NASA to issue an interim report based on more up-to-date findings. NASA, however, decided not to do so. A NASA spokesman said, "We aren't going to put out [another] press release until we have a complete story to tell." The earlier press release was justified, he said, by the high chlorine monoxide levels detected.[94]

NASA WITHDRAWS ITS WARNING

On 30 April 1992, the NASA arctic ozone team officially concluded the arctic project and announced its findings based upon the seven months of data collected. Team leaders declared that despite their earlier fears, an ozone hole had not formed over the Arctic during the previous winter. Nevertheless, they said the threat of an ozone hole would exist each year because of man-made pollutants in the upper atmosphere. They pointed out the record levels of chlorine and other ozone-depleting chemicals in the atmosphere over parts of Europe, Russia, Canada, and the United States. Unusually warm winter air had prevented significant problems. Although there had been a real ozone loss, said Harvard chemist James Anderson, it did not constitute an ozone hole.[95]

The admission of error engendered criticism of NASA, particularly in conservative circles. On the editorial page, the *Washington Times* published a column entitled "NASA Cries Wolf on Ozone." It specifically singled out Michael Kurylo, manager of the Upper Atmosphere Research Program at NASA, for criticism. The editor described Kurylo as having "breathlessly" sounded an alarm. The paper accused NASA of not performing "objective" science and commented, "This is the cry of the apocalyptic, laying the groundwork for a decidedly non-scientific end: public policy." If public policy was its true purpose, the paper said, the strategy had "worked," but it warned NASA against crying "wolf" again.[96] The *Wall Street Journal* also took NASA to task, saying, "The turnaround is another blow to the credibility of the space agency."[97] Having earlier received praise for linking science to policy in the case of Antarctica, NASA now garnered ridicule for wrongly predicting significant depletion of the ozone layer over the Arctic.

Stage 7—Later Implementation

The arctic ozone dispute marked the end of a period in which the ozone issue itself received enormous publicity. Subsequently, the program's implementation moved forward, but in a lesser spotlight. Also, while these events involving ozone were taking place, NASA was undergoing a profound change at the top. In 1989, President Bush appointed Richard Truly NASA Administrator. Because of ongoing disputes between Truly and the White House, President Bush decided in 1992 to remove Truly and appoint Dan Goldin in his place. Prior to his appointment, Goldin had been an executive with TRW (a government contracting firm) and had worked primarily on classified programs with the Department of Defense. Goldin took office on 1 April 1992 with a mandate for change from the White House. In very short order, Goldin made it clear that he would be a very aggressive leader and would align NASA with the new post–Cold War environment that promised flat budgets for the 1990s.

THE GOLDIN TOUCH

Goldin preached that NASA had to operate its programs "faster, better, cheaper." Initially, the emphasis was on using smaller satellites. UARS was an example of the "old NASA," which built billion-dollar satellites. But UARS had already been launched. Hence, Goldin had more impact on the coming programs, represented chiefly by the Earth Observing System. EOS was a comprehensive, advanced monitoring system that would cover all key elements of Earth's system (atmosphere, ocean, biosphere, polar regions) and in the process would also take over the ozone-research work from other NASA satellites.

When he arrived in 1992, Goldin did not like EOS at all. In fact, prior to becoming NASA Administrator and while still at TRW, he had criticized EOS and had drawn the ire of OSSA's leadership. In his view, OSSA leaders had threatened him with loss of possible NASA work unless he refrained from criticism. Now he was the NASA Administrator and in a position to force his agenda on OSSA.[98]

After six months, during which time he used more participatory approaches to induce change, Goldin decided he had waited long enough. In September 1992, he announced a reorganization that removed the highly regarded Associate Administrator of OSSA, Len Fisk, and divided OSSA into three parts.[99] One division would deal with space science, another with life and microgravity science, and the third—with space applications. The applications component was the office called Mission to Planet Earth. Fisk's deputy, Shelby Tilford, was put in charge temporarily, and he, too, eventually left NASA. Goldin took control of Fisk's former domain and began reshaping it into a faster, better, cheaper mode.

REORGANIZING—BUT MAINTAINING OZONE OBSERVATION

EOS was restructured with an eye toward saving money and performing the science through a series of specialized satellites rather than a few, very large satellites. This restructuring had begun prior to Goldin's tenure, propelled by the White House and Congress. But Goldin gave it an even greater push from the Administrator's office. This push meant cutting the program's funding over the remainder of the decade, from $11 billion to less than $8 billion. Cutting the budget inevitably meant the EOS scientists would have to pare down the program's scientific work. As NASA managers and scientists restructured the program, they had to decide which research goals EOS would feature. UARS was in orbit and other existing satellites were capable of being used for ozone research. NASA decided eventually to delay the development of an EOS ozone-related satellite. Satellites emphasizing land or ocean monitoring would come first.

However, NASA made a second decision to maintain ozone observations in the period between UARS and the development of the EOS ozone satellite. The Agency's assumption was the nation and the world needed to know if the Montreal Protocol and its amendments were producing the intended results in mitigating ozone depletion. Also, the nations who had signed the Montreal Protocol needed to know if the substitutes the chemical industry deployed would help stabilize the ozone layer or create a problem themselves. How long would UARS last? When would NASA have the money to launch the EOS ozone satellite? No one could be sure of the answers as the program reviews and restructuring commenced under Goldin's faster, better, cheaper directive. NASA broke up and stretched the EOS program. Its science advisors concluded there had to be an interim system to maintain continuous observations. UARS was not expected to last, in optimal operating conditions, much beyond 1998. The EOS ozone satellite—what would eventually be called Aura—was unlikely to be available until the early 21st century. Monitoring was not as glamorous a form of science as discovery, but for the global environment, it was deemed essential.[100]

THE CLINTON YEARS

In January 1993, Bill Clinton became President and Al Gore his Vice-President. After some hesitancy, Clinton decided to retain Goldin. Those who were interested in environmental issues, including the ozone-depletion issue, saw Gore's arrival as highly positive. He had recently authored a book, *Earth in the Balance*, that showed a deep concern for the same kind of global environmental perspective NASA emphasized. Hence, NASA could be sure its Mission to Planet Earth would survive and possibly be favored by the White House. On the other hand, Clinton constrained the overall NASA budget. The signals from the White House were mixed. Meanwhile, in April, Goldin made another change to the ozone-research program. For some time, Watson had been pushing unsuccessfully for using remotely piloted vehicles, or drones, to keep track of ozone loss in the stratosphere—as complements to satellites. Goldin, with his faster, better, cheaper philosophy, could see how these drones could be useful to NASA's MTPE and also to aeronautics. The aeronautics division of NASA was interested in research to produce such aircraft. Goldin approved a five-year plan to develop research-relevant drones that were capable of lingering in the upper atmosphere longer than other aircraft.[101] This plan was, in effect, a joint venture between MTPE and aeronautics. Unfortunately, it produced no useable vehicles.[102]

The Space Shuttle was also used in connection with the ozone research effort within the same timeframe. On 8 April 1993, the Shuttle *Discovery* blasted into orbit. Its purpose was to gather as much data as possible about the Sun and the Earth's atmosphere. *Discovery* carried a small, retrievable satellite and seven instruments making up the Atmospheric Laboratory for Applications and Science (ATLAS). The mission's aim was to determine how the Sun affected chemical processes in earth's atmosphere, and assess the conditions of the ozone layer over the Northern

Hemisphere, an issue about which there had been much concern and which had caused NASA embarrassment in 1992.[103]

Based upon data from UARS, NASA reported that ozone-destroying chlorine existed longer in the 1992–93 winter than in the previous one. However, scientists reached a consensus that after the year 2000, the outlook for the ozone layer looked better. Even environmentalists were breathing more easily. "The current and projected levels of ozone depletion do not appear to represent a catastrophe," said Michael Oppenheimer of the Environmental Defense Fund.[104]

In August 1993, the *New York Times* editorialized, "Good News, for Once, on Ozone." Citing a recent *Nature* article, it declared that scientists had found contamination of the ozone layer by CFCs to be slowing. The *New York Times* article attributed this fact to the Montreal Protocol and its amendments in 1990 and 1992. The deadline to end CFC production was now 1995, and the policy appeared to be working. If the trend were to continue, scientists could expect the buildup of ozone-destroying chemicals to stop by the year 2000, and subsequently to recede. The atmosphere would then recover. According to the *New York Times*, it seemed as though "science, industry, and government" had agreed on the causes of a problem and then worked to solve it, thus leading the *Times* to proclaim, "there's reason to celebrate."[105]

The apparent success in resolving the problem brought praise for those scientists associated with it, including NASA's Watson and NOAA's Albritton. Both men received awards from the American Association for the Advancement of Science (AAAS).[106] Moreover, Vice-President Gore elevated Watson in October 1993 from NASA to a new position, Associate Director for Environment of the White House Office of Science and Technology Policy (OSTP).

Watson had played a unique role in the ozone saga. He may well have been the right man in the right place at the right time. He accelerated the research, and also played a key role in linking science to policy. His actions, especially those taking place in the international arena, left many of his NASA colleagues wondering how he had the money and stamina for all the international travel involved. Possibly the timing, and NASA's own ambition to develop an environmental mission, created the perfect opportunity for Watson.

OZONE AND CLIMATE CHANGE

Watson's role decreased in importance when the national priority for ozone depletion diminished. His successor, Mike Kurylo, had a very different stage upon which to perform. The spotlight was moving from ozone depletion to climate change in national and international policy. NASA still had to monitor the ozone problem and required, for the long haul, an EOS satellite, but the widespread perception on the part of scientists, the administration, Congress, and the international community was that the main scientific issues of ozone depletion had been resolved. These parties also viewed the related policy issues as having been addressed. The emphasis, in terms of both science and policy, transferred to climate change.

In 1992, President Bush had gone to Rio de Janeiro, Brazil, where leaders of the world's nations gathered to discuss issues of environment and development. The conference did not produce a firm climate-change agreement, at least not one with deadlines and emissions targets. The Bush administration argued that more research was needed to determine the parameters of the problem—if it was really a problem.

Although climate change would become an issue that would keep NASA in the environmental field, it also would move the ozone problem down on the priority list. Ozone monitoring became a more routine, though still important, matter for NASA.

NOAA, a long-term ozone-research partner of NASA, also continued to be involved. While many scientists saw the ozone problem as fading, Susan Solomon of NOAA disagreed. (Solomon led the first ozone expedition to Antarctica in 1986.) Instead of an end, she saw a beginning. "Ozone has begun to enter the climate debate," she said, continuing, "Ozone is a player in the greenhouse effect." Also, she declared that CFC substitutes would have to be monitored by those responsible for enforcing the Protocol. Finally, while the rate of increase was dropping, she believed there would be an increase for some years to come. "The problem is not *close* to getting over," she emphasized.[107]

Ernest Hilsenrath, a scientist at NASA in charge of an ozone monitor aboard the Space Shuttle *Atlantis*, had a different view. He indicated that monitoring the ozone layer required meticulous attention to subtle details of instrument calibration if long-term trends were to be accurately measured. "There aren't any real discoveries," he said. "We haven't found anything new about the Antarctic ozone hole. What we're doing now is providing a baseline for measurements for the future," he continued. "I call it a legacy for environmental investigations in the next century." In the future scientists would be able to use the baseline data (historical data) for comparison to determine the changes.[108]

On 19 December 1994, UARS confirmed with the most conclusive evidence to date that man-made chlorine in the stratosphere, and not some other hypothetical factor, caused the Antarctic ozone hole. Mark Schoeberl, a UARS project scientist, said that UARS data collected for over three years eliminated sea spray, volcanic gases, and other possible natural sources of ozone depletion.[109] While NASA and its allies had held this view for some time, the multiyear UARS global data now made it virtually impossible for anyone to argue that the majority of chlorine in the stratosphere was not from industrially-produced compounds.[110]

IN THE MIDST OF POLITICAL CONFLICT

In 1995, a scientific and political reaction to environmental regulation took place owing, in part, to a Republican majority in both houses of Congress for the first time in decades. The "Republican revolution," as it was called, had as one of its planks the rolling back of unneeded environmental constraints on industry.

Most of the political backlash against regulatory environmentalism was aimed at EPA. However, the negative reaction also affected certain science agencies, NASA included, because of their role in the ozone issue and because of Clinton-Gore support for a tougher stand on climate change. The U.S. Global Change Research Program provided the major thrust for climate-change science, but NASA, with its Mission to Planet Earth, was the largest component of USGCRP. Robert Walker (Republican, PA), chairman of the House Committee on Science, questioned the validity of MTPE science and whether it was being used to justify pre-existing political views. He was targeting EOS, primarily, but his attack (which included the threat of a major budget cut) was indicative of the conservative backlash against ozone regulation.

The *Washington Times* conducted a steady drumbeat of criticism of both the Montreal Protocol and the domestic regulation that came from it. The paper indicted NASA regularly for its "political" science. Another conservative publication, *21st Century*, took issue with NASA's December 1994 statement that UARS data had proved conclusively that man-made chlorine and not natural causes, such as volcanoes, was responsible for the Antarctic Ozone hole. It accused NASA of trying to close scientific debate prematurely. NASA officials regarded such comments as distortions of the scientific issues and debate, especially with respect to NASA's role. Nevertheless, these critical comments caused difficulties in the Agency's political relationships.[111]

In July 1995, NASA announced that the ozone problem had reached a different stage: that levels of ozone-depleting chemicals were falling. Dr. Robert Harris, head of MTPE's science division, declared, "We've already seen slowing increases in chlorofluorocarbon and other ozone-depleting chemicals. Now, for the first time, we've actually measured a decrease in one of these chemicals."[112]

In September 1995, in an effort to roll back ozone regulation, conservative Republicans in Congress moved to postpone the phasing out of certain ozone-destroying chemicals. Representative John Doolittle (Republican, CA) introduced a bill to delay the CFC ban from 1996 to 2000.

The House held hearings on the bill and Watson, now employed by the White House OSTP, testified in opposition. He said harmful ultraviolet radiation reaching Earth's surface "will, not may" have adverse consequences for human health and ecological systems. He also said there was "absolutely no doubt" that the problem was caused by human activities, not natural processes. His testimony, he remarked, represented the views of "the very, very large majority" of scientists working on the problem.[113]

However, Sallie Baliunas, a biophysicist representing the George C. Marshall Institute (a Washington-based research organization) and Fred Singer, professor emeritus of environmental sciences at the University of Virginia, challenged Watson's view on the basis of their research.[114]

The legislation did not become law. Moreover, legislative efforts to cut back NASA research also largely failed, and the following year, a congressionally requested review by the National Research Council strongly endorsed EOS. Countering criticism that EOS was politically driven, the NRC said science drove the program and the program should be supported in the future.

As these external debates were taking place, internally NASA had to cope with extremely constrained funding—the Clinton Administration continued to cut the Agency's budget year after year. NASA and its science advisors held to their view that continuous monitoring of ozone depletion was essential. Worried that the EOS ozone satellite's launch was being pushed farther into the future by Administration and congressional budget cuts, and that UARS was aging, NASA had decided to mount an interim sequence of ozone monitoring equipment that could be justified in faster, better, cheaper terms and prevent a data-collection gap between UARS's demise and the EOS ozone satellite launch. This system involved small satellites and, in some cases, meant putting U.S. sensors on foreign satellites.[115]

STAGE 8—INSTITUTIONALIZATION

In June 1996, NASA produced an ozone–depletion report that showed concentrations of ozone-depleting chemicals were beginning to level off. The policy was working, and nations and industry were following the agreements. The need for continuous monitoring gave NASA a potentially long-term mission in ozone observation. It also gave NASA a long-term problem: How was an agency with an R&D mission supposed to incorporate and fund relatively routine monitoring?

TOMS EARTH PROBE

In addition to UARS (the most comprehensive existing ozone-related spacecraft), other spacecraft began to be deployed—ones that NASA had authorized for the interval between UARS and the EOS satellite launch. In July 1996, NASA launched the TOMS Earth Probe. In September 1996, the Japanese launched the Advanced Earth Observing Satellite, which used a TOMS instrument to map ozone depletion daily. These activities extended the series of TOMS observations that had begun with Nimbus in 1978.

In May 1997, an enhanced NASA ER-2 conducted its first operational mission. Kurylo, manager of the Upper Atmosphere Research Program, said the high-flying airplane was critical—it enabled NASA to reach the intermediate region between where aerosol-particle-driven processes are measured by standard aircraft-based sensors and where gas-phase processes are monitored by orbiting satellites.[116]

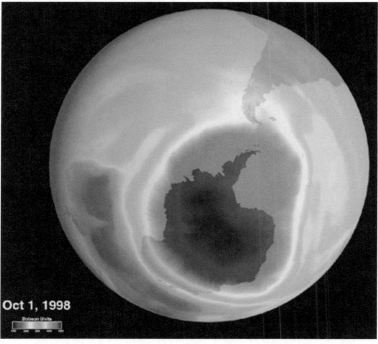

Southern Hemisphere ozone hole, 1998. (Source: http://grin.hq.nasa.gov/ABSTRACT/GPN-2002-000117)

In early August, the Space Shuttle *Discovery* rocketed into orbit and released a German satellite to study Earth's ozone layer. *Discovery* returned later in the month bringing the satellite and its data home. The primary goal of this 86th Shuttle mission was to launch and retrieve the satellite after it had made nine days of independent atmospheric observations. The satellite's measurements supported UARS findings.

The Japanese ADEOS ended its life late in 1997. In December, NASA boosted its Total Ozone Mapping System (TOMS) Earth Probe satellite into a higher orbit, a position from which it could widen its coverage. During 1998, scientists from NASA's Goddard Institute for Space Studies and the Center for Climate Systems Research at Columbia University warned that the depletion of the ozone layer would get worse, not better, in the future due to the impact of carbon dioxide and other greenhouse gases. In the wake of the global controls on ozone-destroying compounds, many observers had expected the annual Antarctic ozone hole to diminish in size, along with the more modest Arctic ozone losses. In April, the NASA-Columbia team predicted that during the next few decades greenhouse gases would trigger a springtime ozone hole over the Arctic, much like the ones seen over Antarctica.

Although there was scientific debate over this prediction, Richard Kerr, writing in *Science* magazine, said that the NASA-Columbia model "enhanced" the awareness of two great human alterations of the atmosphere—greenhouse warming and ozone depletion—which were indeed interdependent."[117] Between mid-August and early October 1998, NASA and NOAA found that the antarctic ozone hole was setting a new record for size. The NASA-Columbia prediction and the NASA-NOAA findings pointed out the need for constant surveillance of the ozone problem.[118]

THREAT OF DATA INTERRUPTION

Toward the end of 1998 and into 1999, a number of scientists became worried about NASA's ability to have a sufficient flow of ozone data. UARS was still functioning but was widely viewed as not having much of useful lifetime left. The NASA TOMS Earth Probe suffered a three-week malfunction near year's end. It resumed collecting data in January 1999, but the satellite was past its designed two-year life span and was operating on a back-up transmitter, the primary one having failed the previous April.

The worry of scientists increased when NASA lost the host satellite for its next TOMS instrument. The sensor was supposed to fly aboard a Russian meteorological satellite that was to be launched in 2000 via a Soyuz rocket. The Russians informed NASA that the satellite's launch had been postponed indefinitely due to Russia's financial problems.

In November 1999, NASA scientists joined researchers from Europe, Russia, Canada, and Japan in the largest field-measurement project in history to assess ozone amounts and changes in the wintertime arctic upper atmosphere. The project was designed to run from November 1999 through March 2000. It used satellites, airplanes, heavy-lift and small balloons, and ground-based instruments. The U.S. cost was $20 million. Other nations contributed a total of $10 million.

The project required a tremendous concentration of science instruments to be placed in one of the coldest regions of the arctic stratosphere. The satellites involved included NASA's TOMS Earth

NASA and the Environment

Probe, UARS, and the Earth Radiation Budget Satellite; the U.S. Air Force's Midcourse Space Experiment; the European Space Agency's Earth Remote Sensing Satellite; and France's Satellite Pour l'Observation de la Terre (SPOT 4).

This project was designed to shed light on whether a new ozone hole was forming above the Arctic. More than 150 scientists and technicians traveled northward to participate in the international effort.[119] In April 2000, scientists working on the arctic project reported that the ozone layer was not recovering as fast as they had hoped. Again, some proposed that global warming was offsetting gains in ozone-layer recovery. The new research showed that the ozone layer had thinned to record levels over the Arctic and that the thinning would probably worsen over the next decade. Ozone-layer thinning over the Arctic was much more dangerous to Americans, because the weakening might migrate south in the spring. Such a weakening would mean northern U.S. cities could be in danger.

THE UARS CONTROVERSY

In August 2000, scientists learned that NASA had decided to exclude UARS from its FY 2002 budget in order to help make funds available for new programs, especially Aura, (the advanced EOS atmospheric satellite that would take the place of UARS and the Agency's other ozone-related satellites). NASA's role, one could argue, was to innovate new satellite technology, not operate old satellites indefinitely. With tight budgets, the old equipment had to make room for the new. Hence, NASA did not foresee needing funds for future UARS operations.

In September, however, NASA announced that it had detected an antarctic ozone hole three times larger than the entire U.S. landmass—the largest hole ever observed. Kurylo declared, "These observations reinforce concerns about the frailty of Earth's ozone layer. Although production of ozone-destroying gases has been curtailed under international agreements, concentrations of the gases in the stratosphere are only now reaching their peak. Due to their long persistence in the atmosphere, it will be many decades before the ozone hole is no longer an annual occurrence."[120]

The continuing ozone problem, apparently complicated by greenhouse warming, made NASA's decision to phase out UARS increasingly controversial among scientists in academia and within NASA.

THE GEORGE W. BUSH ADMINISTRATION

In January 2001, the George W. Bush administration came to power. It adjusted the NASA budget it had inherited and shifted money among various programs. The spending for Earth Sciences Enterprise, formerly called Mission to Planet Earth until 1998, went down 13.9 percent in spending.[121] This budget cut strengthened the view within NASA that UARS should not be kept in space any longer than necessary.

NASA scheduled UARS to cease operating on 30 September 2001. The Agency said it cost $10 million annually to maintain the Earth-orbiting satellite. "We don't have the funding to continue the work of the satellite, so we are going to decommission it," David Steitz, a NASA spokesman, declared.

A number of scientists who worked on the ozone-depletion problem were not happy. "How can NASA turn off a satellite?" they asked. "We have planned the decommissioning of this for years," Steitz responded. Originally, UARS had been projected to last three years. Instead, it had lasted 10. Aura would replace it, in any event, the Agency said. Unfortunately, Aura was not moving very quickly toward deployment and many scientists feared there would be a gap in the kind of comprehensive data UARS had provided.

In the spring of 2001, NASA's Goddard Space Flight Center had submitted a proposal to NASA Headquarters in a last-ditch effort to save UARS. It was to no avail. "It's a $1 billion asset we're throwing down the drain because we can't come up with a couple of million to keep it running," charged Mark Schoeberl, the mission's former project scientist at Goddard. "Sorry guys," said Steitz, "but it's over. We can't afford to continue to feed it and we have other priorities with new technologies."[122]

NASA officials in Washington, DC, argued that there was no danger of an immediate loss of the Agency's ozone-monitoring capability, because the TOMS Earth Probe satellite was still on orbit and in reasonably good health. Critics pointed out that this spacecraft's ozone-mapping instrument was already three years past its planned lifetime and was not as strong as it once was.

NASA launched a supplementary satellite in September 2001, the $35-million Total Ozone Mapping Spectrometer (dubbed QuikTOMS). This satellite, however, was destroyed in a launch failure. Richard McPeters, NASA's project manager for QuikTOMS, said the loss "is really throwing ozone monitoring programs into disarray."

Following the loss of QuikTOMS, Ghassem Asrar, associate administrator for the Earth Science Enterprise, formed a team to deal with various aspects of the data gap-UARS controversy. He said that keeping UARS as a backup to the TOMS Earth Probe was possible. Postponing UARS' retirement would be an "insurance policy" in case the TOMS Earth Probe failed before Aura was launched.[123]

NASA scrambled and found the money to keep UARS going. In March 2002, the European Space Agency launched Envisat, with three ozone-measuring instruments. This satellite provided an extra measure of security for continuous ozone monitoring. Aura, now scheduled for launch in 2004, would be the ultimate salvation.

LOOKING AHEAD

In September 2002, UNEP/WMO's Scientific Assessment of Ozone Depletion found that "restraints on production of ozone-destroying chemicals such as CFCs are having the intended effect." Paul Newman, a NASA atmospheric physicist and the assessment's coauthor, said that "by 2010, we could see five to six years when the hole looks consistently smaller than during the past five years."

Shortly afterward, NASA and NOAA reported that the ozone hole was the smallest it had been since 1988. Newman attributed the decrease of ozone loss to unusual stratospheric weather patterns, specifically warmer-than-normal temperatures of Antarctica's polar vortex. European scientists disagreed, attributing the decrease to CFC reduction measures taken in the 1980s.[124]

In August 2003, a team of researchers from NASA, the University of Alabama at Huntsville (UAH), Georgia Institute of Technology, the University of Wisconsin, and Hampton University reported that the rate of ozone depletion was again decreasing. Moreover, the rate of increase of upper-stratospheric chlorine, which destroys ozone, was also slowing. This evidence provided the clearest indication yet that the Montreal Protocol and subsequent restrictions were having the intended effect.

"This is the beginning of recovery of the ozone layer," declared Michael Newchurch, an atmospheric chemist at UAH and the lead researcher on the ozone-trend-analysis team. The group used data from three NASA satellites and three international ground stations. It found ozone depletion had definitely slowed since 1997. To be sure, the problem of ozone depletion continued and the Newchurch data were not unanimously accepted.[125] Nevertheless, Newchurch predicted there would be improvement in the coming years. "We had a monumental problem of global scale that we have started to solve," he stated."[126]

Conclusion

The ozone-depletion case is almost universally seen as a success story in the link between science and policy. In the environmental field, such success stories are few. Hence, it is worth considering why science and policy worked together in this instance. The focus of the preceding narrative has been on NASA's role in the overall link. There are probably 1,000 heroes in any successful public-policy case, and that, no doubt, is true in this one. NOAA, NSF, environmentalists, and even industry can share in the credit. Nevertheless, a key factor in the ozone-depletion issue was that there had to be someone in charge from the science side of the science-policy equation. There was a de facto "lead" agency—NASA.

There is much interest on the part of policy-makers in mechanisms for constructive scientific action in environmental policy problems that also cross agency and international boundaries. Hence, this particular case study can teach important lessons about the role of a lead agency. Below is a list of significant factors and events:

1. NASA had legislative authority. NASA obtained lead agency status by virtue of legislation passed in 1975 and 1977. This legislative authority gave NASA influence in asserting its claims in interagency relations. NASA had the legitimacy to pursue research *and* to make assessments of what this research meant for policy.

2. NASA's role evolved with changes in public policy. That is, NASA adapted its ozone-depletion program as policy needs changed. It was responsive to its environment, and this, in turn, helped it influence policy.

3. In the late 1970s and early 1980s, NASA built a "science base" of knowledge. At key points in time (in the mid and latter 1980s and early 1990s), it mobilized science resources, organized expeditions that accelerated policy-relevant scientific research, and helped find the "smoking gun" linking CFCs to ozone depletion. NASA also speeded the assessment of scientific findings for policy-makers through mechanisms that enabled scientists to build consensus about what the data indicated in terms of trends.

4. NASA was highly "ecumenical"[127] in its strategies by enlisting other agencies (especially NOAA) and industry groups. This participative strategy to reach scientists outside the United States helped to bring entities with different interests into agreement.

5. NASA's role in the ozone-depletion debate illuminates one very complex issue: the timing of scientific communication to policy-makers. Science usually moves at a pace quite different from policy. Political forces drive policy. The search for truth helps drive science, and that requires a careful checking of facts, peer review, and publication in scientific literature. But the policy debate may require science to speak up early in order to take advantage of policy-making opportunities. When scientists speak out and issue assessments, these opinions can take the shape of warnings and such warnings can influence policy. Scientists are also, thereby, participating in the politics that shape policy. When policy outcomes are widely seen as positive, as with the Montreal Protocol, such warnings are, in hindsight, regarded as useful. But when policy outcomes are contested

or later events show the warnings to have been premature (as was the situation with the NASA warning of dangerous arctic-ozone depletion in 1992), the lead agency gets burned. Lead-agency status in linking science and policy has its dangers as well as its advantages and must be exercised with care.

6. The ozone-depletion case highlights the long-term nature of the lead agency role vis-à-vis science/policy relations. Since the Montreal Protocol, the primary policy role of science has been to monitor the degree to which the Protocol and its amendments have been effective. This monitoring role is long-term and not particularly glamorous. When budgets are tight, there is a temptation to cut monitoring, especially in an R&D agency geared to the future. Nevertheless, monitoring is critical and complicated by the inevitable need to replace aging technology. NASA made a major decision in the mid-1990s to maintain monitoring in spite of extreme turbulence within the Agency and relentless budget pressure from the outside.

7. One of the reasons NASA was successful as a lead agency was that it not only had legislative authority, but it also had the requisite financial resources and administrative and political support to play that role. It also had competent personnel, including a very skillful leader (Robert Watson), at a pivotal point in history. Watson was an able alliance-builder, manager, and science adviser; but he would not have succeeded without the support of his top management and an urgent political setting. Edward Parson, a Harvard policy researcher who has written on the ozone issue, has called Watson's strategy "the authoritative monopoly strategy. . . . It entails conducting an assessment by assembling an international near-monopoly of relevant expertise to present an authoritative statement of present knowledge."[128]

8. Finally, the ozone-depletion case reveals how an agency's mission evolves over time. NASA has become an increasingly important science agency in global environmental affairs, including climate change. This process of mission innovation began with ozone depletion. NASA built on the ozone-depletion issue in the "amplification" stage of policy development. It used this issue to help reinvent itself, promote the Earth Observing System, and propel itself onto the stage of climate change.

Endnotes

[1] The author thanks NASA for support under which this research was performed. The NASA contract was W-92701. The author also thanks NSF (grant number SES-0114689) for general support on the change process at NASA. The author acknowledges the research assistance of Anne Hardenbergh of the Maxwell School's graduate program in International Relations, Larry Epstein of Maxwell's Public Administration graduate program, and Lisa Van Arsdale of Maxwell's Public Administration graduate program. The author also thanks various reviewers, some named in the notes, others anonymous, who provided detailed information on the sequence of events. Finally, thanks also go to Cathy Wilson, Paula Kephart, Steve Bradley, Jeffrey McLean, and Henry Spencer of the NASA Headquarters Printing and Design Office for their expert help, as well as to Stephen Garber of the NASA History Division for his knowledgeable assistance.

[2] Richard Benedick, *Ozone Diplomacy* (Cambridge, MA: Harvard, 1991); Stephen Andersen and K. M. Sarma, *Protecting the Ozone Layer: the United Nations History* (London: UNEP, Earthscan, 2002); Maureen Christie, *The Ozone Layer: A Philosophy of Science Perspective* (Cambridge, England: Cambridge University Press, 2000); Edward Parson, *Protecting the Ozone Layer: Science and Strategy* (New York: Oxford, 2003).

[3] Study of Critical Environmental Problems (SCEP) and William H. Matthews, eds., *Man's Impact on the Global Environment: Report of the Study of Critical Environmental Problems* (Cambridge, MA: MIT, 1970). Cited in Erik Conway, "High Speed Dreams," draft report to NASA History Division (1 July 2002): 295.

[4] Conway, pp. 296–300. See also Harold Johnston, "Reduction of Stratospheric Ozone by Nitrogen Oxide Catalysts from Supersonic Transport Exhaust," *Science* (6 August 1971): 517–522.

[5] Senator Clinton Anderson to James Fletcher, 10 June 1971, Washington National Records Center, Suitland, MD. Acc#255-77-0677, Box 52, file "NASA's Advanced SST Program, Catalytic Destruction of Ozone." Cited in Conway, p. 300.

[6] Conway, pp. 300–302.

[7] R. J. Cicerone et al., "Assessment of Possible Environmental Effects of Space Shuttle Operations," NASA CR-129003 (3 June 1973).

[8] Conway, pp. 303–304. There was at least one other report, by Lockheed, with similar views. Information provided by personal communication from Richard Stolarski of NASA Goddard, 16 December 2003. Also, even though SST was killed in 1971, Langley Research Center continued a Supersonic Cruise Aircraft Research Program until 1981.

[9] M. J. Molina and F. S. Rowland, "Stratospheric Sink for Chlorofluoromethanes: Chlorine Atom Catalyzed Destruction of Ozone," *Nature* 249 (28 June 1974): 810–812.

[10] Conway, pp. 303–304.

[11] Conversation between John Naugle and Ed Todd, memo for record, 10 December 1974, Ozone file, NASA History Division, NASA Headquarters, Washington, DC.

[12] Ibid.

[13] Ibid.

[14] S.851. Cited in Congressional Record-Senate (26 February 1975), S2641.

[15] Ibid.

[16] Stolarski, 16 December 2003.

[17] "NASA's Chief Urges Delay on Aerosol Ban," *Washington Post* (16 September 1975): A3. Conway, p. 305.

[18] National Research Council, *Response to the Ozone Protection Section of the Clean Air Act Amendment of 1977: An Interim Report.* (Washington, DC: National Academies of Sciences, 1977).

[19] The Clean Air Act as amended August 1977 and July 1980 (Washington, DC: Government Printing Office, 1980), 96th Cong. 2d Sess. There were numerous scientific reports by various bodies in the 1970s. The decade ended with one by the National Academy of Sciences. See National Academy of Sciences, National Research Council, *Protection Against Depletion of Stratospheric Ozone by Chlorofluorocarbons* (Washington, DC: National Academy of Sciences, 1979).

[20] A useful study of ozone politics is by Sharon Roan, *Ozone Crisis* (New York: John Wiley & Sons, Inc., 1989). National Research Council, *Response to Ozone Protection*.

[21] One early report was R. D. Hudson and E. I. Reed (eds.), *The Stratosphere: Present and Future* (National Aeronautics and Space Administration, RP-1049, 1979). NASA had hoped to launch UARS in 1984. Stolarski, 16 December 2003.

[22] Sharon Roan, *Ozone Crisis* (New York: John Wiley, 1989), p. 98.

[23] W. Henry Lambright, "Administrative Entrepreneurship and Space Technology: The Ups and Downs of 'Mission to Planet Earth'," *Public Administration Review* Vol. 54, no. 2 (March/April 1994): 99.

[24] Ibid.

[25] Personal communication from Jack Kaye, 16 December 2003.

[26] Nick Sundt, "Robert T. Watson: Bridging Science and Policy," *Global Change* (Electronic Edition) (May 1996).

[27] Interview with Jack Kaye, 18 October 2002.

[28] Roan, p. 159.

[29] Karen Litfin, *Ozone Discourses* (New York: Columbia, 1994), p. 82.

[30] Ibid.

[31] Ibid., p. 112.

[32] Ibid., p. 82.

[33] Ibid., p. 83.

[34] Ibid., p. 73.

[35] Ibid., pp. 75–77.

[36] J.C. Farman, et al., "Large Losses of Total Ozone in Antarctica Reveal C10x/NOx Interaction," *Nature* 315 (16 May 1985): 207–210.

[37] Stolarski, 16 December 2003.

[38] Interview with Jack Kaye, 18 October 2002.

[39] Litfin, p. 83. Roan, p. 142. W. Henry Lambright, "NASA, Ozone, and Policy-Relevant Science," *Research Policy* 24 (1995): 752.

[40] Interview with Jack Kaye, 18 October 2002.

[41] Lambright, "Entrepreneurship and Space Technology," p. 99.

[42] Lambright, "NASA, Ozone, and Policy-Relevant Science," p. 753.

[43] Ibid. Roan, p. 172. The data pointed to changes in partitioning of chlorine and nitrogen compounds, and large concentrations of "acute" forms of chlorine in the lower atmosphere where models and previous observations suggested that chlorine would be present in less chemically active forms. Personal communication from Jack Kaye, 16 December 2003.

[44] Anonymous reviewer comment. (All anonymous reviewer comments came from NASA review of this manuscript. The comments are on the original manuscript but remain private due to anonymity of the peer review process.) See also M. R. Schoeberl and A. J. Krueger (guest eds.)"Overview of the Antarctic Ozone Depletion" *Geophysical Research Letters* 13, no. 12 (November Supplement 1986): 1191–1362.

[45] Litfin, p. 104.

46 Ibid., pp. 105–106. "Through Rose-Colored Sunglasses," *New York Times* (31 May 1987): 28.

47 Lambright, "NASA, Ozone, and Policy-Relevant Science," p. 753.

48 Personal communication from Jack Kaye, 16 December 2003.

49 Lambright, "NASA, Ozone, and Policy-Relevant Science," p. 753. Interview with Robert Watson, 22 June 1993.

50 Anonymous reviewer comment.

51 Anonymous reviewer comment.

52 Roan, p. 183. This was not the first large airborne campaign to study stratospheric chemistry. There had been two previous ones in the 1980s. This experience helped NASA in the present instance. Personal communication from Jack Kaye, 16 December 2003.

53 Sally Ride, *Leadership and America's Future in Space* (Washington, DC: NASA, 1987).

54 Litfin, p. 85.

55 Personal communication from Jack Kaye, 16 December 2003.

56 Lambright, "NASA, Ozone, and Policy-Relevant Science," p. 754.

57 Personal communication from Jack Kaye, 16 December 2003. See also: "The Airborne Antarctic Ozone Expedition (AAOE), Part 1," *Journal of Geophysical Research* 94, no. D9 (30 August 1989): 11179–11737, especially A. F. Tuck et al., 'The Planning and Execution of ER-2 and DC-8 Aircraft Flights Over Antarctica, August and September 1987,' pp. 11181–11222, and "The Airborne Antarctic Ozone Expedition (AAOE), Part 2," *Journal of Geophysical Research* 94, no. D14 (30 November 1989): 16437–16857.

58 Anonymous reviewer comment.

59 Interview with Dan Albritton, 15 June 1993.

60 For a comprehensive treatment of the diplomatic story, see R. E. Benedick, *Ozone Diplomacy*, (Cambridge, MA: Harvard, 1991).

61 Lambright, "NASA, Ozone, and Policy-Relevant Science," p. 754.

62 Personal communication from Jack Kaye, 16 December 2003.

63 Lambright, "NASA, Ozone, and Policy-Relevant Science," p. 755.

64 Anonymous reviewer comment.

65 Lambright, "NASA, Ozone, and Policy-Relevant Science," p. 755.

[66] Ibid.

[67] Malcomb Browne, "New Ozone Threat: Scientists Fear Layer is Eroding at North Pole," *New York Times* (11 October 1988):11. Once scientists proposed that the polar stratospheric clouds (PSCs) played an important role in repartitioning the chlorine in the Antarctic stratosphere from inactive to active forms, scientists were able to look at more satellite data to understand the distribution of the PSCs. In particular, scientists from the NASA Langley Research Center were able to look at PSC data from the instruments aboard the Nimbus 7 satellite and characterize the distribution of PSCs over the Antarctic and Arctic. This helped provide input to models attempting to simulate the PSC-induced chemistry. Also, lab work done several places (most notably at the University of Minnesota and the NOAA Aeronomy Lab) helped scientists understand which chemicals were likely responsible for forming the PSCs—it was the existence of a broadly-based laboratory program that enabled scientists to quickly focus on what might be responsible for PSC formation and their impacts on chemistry. Personal communication from Richard Stolarski, 16 December 2003.

[68] Norman Vig and Michael Kraft, *Environmental Policy: New Directions for the 21st Century*, (Washington, DC: Congressional Quarterly Press, 2003), p. 306.

[69] Philip Shabecoff, "Arctic Expedition Finds Chemical Threat to Ozone," *New York Times* (18 February 1989): 1,9.

[70] Anonymous reviewer comment.

[71] Craig Whitney, "Ozone Talks End Without Deadline," *New York Times* (8 March 1989): A12.

[72] "Ozone Hole's Reappearance Linked to Chlorofluorocarbons," *Aviation Week and Space Technology* (30 October 1989): 28.

[73] William Booth, "Hole in Ozone Layer Found at North Pole, Too," *Washington Post* (16 March 1990): A10.

[74] Malcolm Browne, "Ozone Hole Reopens Over Antarctica," *New York Times* (12 October 1990): A8.

[75] Douglas Isbell, "UARS Seen as Earth Observing System's Dress Rehearsal," *Space News* (9–15 September 1991): 24 .

[76] Liz Hunt, "'Mission to Planet Earth' Commences as Shuttle Carries Satellite Into Orbit," *Washington Post* (13 September 1991): A3.

[77] Douglas Isbell, "UARS Launches Earth Study," *Space News* (16–22 September 1991): 24.

[78] Personal communication from Jack Kaye, 16 December 2003.

[79] Micah Morrison, "The Ozone Scare," *Insight, Washington Times* (22 March 1992): 15–23.

[80] Ibid.

[81] Anonymous reviewer comment.

[82] Paul Hoversten, "Suspected Northern Ozone Hole 'Frightening'," *USA Today* (4 February 1992): A1. Paul Hoversten, "Scientists Sound Ozone Alarm," *USA Today* (4 February 1992): A3.

[83] Ibid.

[84] *Time* Vol. 139, no. 7 (17 February 1992): cover page.

[85] Michael Weisskopf, "U.S. May Seek to Hasten Action to Protect Ozone," *Washington Post* (6 February 1992): A3.

[86] Philip Hilts, "Senate Backs Faster Protection of Ozone Layer as Bush Relents," *New York Times* (7 February 1992): A1.

[87] Keith Schneider, "Bush Orders End to Manufacture of Ozone-Harming Agents by '96," *New York Times* (13 February 1992): B16.

[88] Michael Weisskopf, "Ozone Depletion Tied to Infectious Diseases; UN Report Could Prompt Bid to Speed Phaseout of Chlorofluorocarbons," *Washington Post* (8 February 1992): A-10.

[89] M. D. Lemonick, "The Ozone Vanishes," *Time* Vol. 139, no. 7 (17 February 1992): 60-63. See also Micah Morrison, "The Ozone Scare," *Insight, Washington Times* (22 March 1992): 15–23.

[90] Ronald Bailey, "Oops...There Goes That Ozone Hole!," *Washington Times* (1 March 1992): 11.

[91] Morrison, "The Ozone Scare." There was an ongoing and intense scientific debate about the migration of ozone loss from the vortex to the mid-latitudes at the time of the press conference (3 February 1992). The factors that influenced this spread, or inhibited it, were better understood later when the second Arctic expedition ended. Anonymous reviewer comment.

[92] Ibid. The situation in the stratosphere was complicated by a very large volcanic eruption of Mt. Pinatubo in the Philippines in 1991. The sulfur dioxide injected into the stratosphere by this volcano led to formation of significant amounts of stratospheric aerosols. Since the lifetime of these aerosols is on the order of one year, it took several years for the amounts to decay. The aerosols had the potential to catalyze some of the same kinds of chemical reactions that PSCs did in the Arctic. The best way to track the growth and decay of the stratospheric aerosols was seen in NASA's second Stratospheric Aerosol and Gas Experiment (SAGE II) launched in 1984. UARS also had an instrument to help in the study of aerosol distribution and properties. Personal communication from Richard Stolarski, 16 December 2003. An anonymous reviewer noted that "Hoffman's comment applied to a single, unusual event, the presence of volcanic aerosols that year. Aerosols affect the reaction in ways they didn't know, and they could have caused the greater-than-usual chlorine abundance in 1991. But it doesn't impugn the original thesis of the prior years' experiments . . . UARS data, in fact, disproved the volcanic thesis later. . . ."

[93] Anonymous reviewer comment.

[94] Morrison, "The Ozone Scare."

[95] Warren Leary, "Scientists Say Warm Winter Prevented Arctic Ozone Hole," *New York Times* (1 May 1992): A12.

[96] "NASA Cries Wolf on Ozone," *Washington Times* (7 May 1992): 62.

[97] Bob Davis, "Hole in Ozone Didn't Develop, NASA Reports," *Wall Street Journal* (1 May 1992): B12.

[98] Stephanie Roy, "The Origin of the Smaller, Faster, Cheaper Approach to NASA's Solar System Exploration," *Space Policy* 14 (1998): 166.

[99] At first, there were two parts, space science and applications (called Mission to Planet Earth). The third, life and microgravity science, came not long afterward.

[100] Ghassem Asrar interview, Washington, DC, 16 December 2003.

[101] Gary Taubes, "NASA Launches a 5-Year Plan to Clone Drones," *Science* (16 April 1993): 286.

[102] Anonymous reviewer comment.

[103] William Harwood, "Shuttle Lifts Off on Mission to Study Ozone," *Washington Post* (8 April 1993): A22. This ATLAS flight was the second in a series (the first one was launched 24 March 1992). This second flight received a fair amount of attention because of what was going on with the ozone layer at the time. The Agency was even willing to do a launch at night (which it prefers not to) in order to provide the best possible view of the Northern Hemisphere high latitudes during spring. ATLAS was supposed to be a series of 10 missions, although it ended up being truncated after 3 (the last one was in 1994). One of the goals of the ATLAS missions was to help provide calibration information for UARS, as there were some comparable instruments aboard ATLAS that could be brought back into the lab and tested after flight, so calibrations could be very well established. The Shuttle also was used to fly the Shuttle Solar Backscatter Ultraviolet (SSBUV) instrument, which provided calibration information for a series of ozone-measuring satellite instruments that flew as part of NOAA's operational meteorological satellite system (these are SSBUV/2 instruments). The SSBUV instrument flew eight times over a period of years. The key instrument aboard ATLAS—the Atmospheric Trace Molecules Spectroscopy Experiment (ATMOS) instrument from the Jet Propulsion Laboratory—also flew aboard the Spacelab 3 mission in 1985. Personal communication from Richard Stolarski, 16 December 2003.

[104] Boyce Rensberger, "After 2000, Outlook for the Ozone Layer Looks Good," *Washington Post* (15 April 1993): A1.

[105] "Good News, for Once, on Ozone," Editorial, *New York Times* (30 August 1993): A16.

[106] NASA, "AAAS to Honor NASA Scientist for Ozone Research," news release (11 February 1993).

[107] Carl Posey, "Ozone Forecast: Partly Cloudy," *Air and Space* (October/November 1994): 34.

[108]"Astronauts Again Seeking Clues About Dwindling Ozone Shield," *New York Times* (10 November 1994): PA 32.

[109]NASA, "NASA's UARS Confirms CFCs Caused Antarctic Ozone Hole," news release (19 December 1994).

[110]Personal communication from Richard Stolarski, 16 December 2003.

[111]Anonymous reviewer comment.

[112]NASA, "NASA Funded Research Sees Fall of Ozone-Depleting Chemicals," news release (20 July 1995).

[113]Committee on Science, Status of the Stratospheric Ozone Layer: Hearing before the Subcommittee on Energy and the Environment, US House of Representatives, 104th Congress, 1st Sess., September 20th, 1995. See also William K. Stevens, "GOP Bills Aim to Delay Ban on Chemicals in Ozone Dispute," *New York Times* (21 September 1995): A20.

[114]William K. Stevens, "GOP Bills Aim to Delay Ban on Chemicals in Ozone Dispute," *New York Times* (21 September 1995): A20.

[115]Ghassem Asrar interview, Washington, DC, 16 December 2003.

[116]NASA, "NASA Earth Science Aircraft Soars to New Heights," news release (13 May 1997).

[117]Richard Kerr, "Ozone Losses, Greenhouse Gases Linked," *Science* 280 (10 April 1998): 39.

[118]Warren Ferster, "Officials Say Ozone Needs Constant Vigil," *Space News* (12–18 October 1998): 12.

[119]Brian Berger, "Key Component Left Out of NASA Ozone Research Project," *Space News* (6 December 1999): 8.

[120]NASA, "Largest-Ever Ozone Hole Observed Over Antarctica," news release (17 September 2000).

[121]The name of the division, Mission to Planet Earth, was changed to Earth Science Enterprise in 1998.

[122]Andrew Bridges, "NASA Terminates Mission that Measured Ozone Hole," Associated Press Newswire (23 August 2001).

[123]Brian Berger, "QuikTOMS Loss Limits NASA's Ozone Monitoring Options," *Space News* (1 October 2001):15.

[124]Pierre Sparaco, "Nations Disagree on Reasons for Shrinking Ozone Hole," *Aviation Week and Space Technology* (11 November 2002): 44.

[125] The author thanks an anonymous reviewer for the suggested material.

[126] D. E. Heath, ed., "Ozone Layer on the Road to Recovery," *Environment* (October 2003): 4–5. Some scientists had problems with the perception that ozone was increasing. Newchurch's data did not show this, and he tried to be clear that what he saw was a slowing of the decrease—a flattening out of the loss rate, which would be a necessary precursor to recovery. Personal communication from Jack Kaye, 16 December 2003.

[127] This is the word Albritton used in describing Watson's strategy in the Montreal Protocol era. Interview with Albritton, 15 June 1993.

[128] Parson, pp. 266–267.

About the Author

W. Henry Lambright is professor of public administration and political science and director at the Center for Environmental Policy and Administration, the Maxwell School, Syracuse University. He is the author of a number of books and publications on public administration and science policy. His works on space policy include *Powering Apollo: James E. Webb of NASA* (Johns Hopkins University Press, 1995), *Space Policy in the Twenty-First Century* (Johns Hopkins University Press, 2003), and *Transforming Government: Dan Goldin and the Remaking of NASA* (PricewaterhouseCoopers, 2001).

NASA Monographs in Aerospace History Series

All monographs except the first one are available by sending a self-addressed 9-by-12-inch envelope for each monograph with appropriate postage for 15 ounces to the NASA History Division, Room CO72, Washington, DC 20546. A complete listing of all NASA History Series publications is available at http://history.nasa.gov/series95.html on the World Wide Web. In addition, a number of monographs and other History Series publications are available online from the same URL.

Launius, Roger D., and Aaron K. Gillette, compilers. *Toward a History of the Space Shuttle: An Annotated Bibliography.* Monographs in Aerospace History, No. 1, 1992. Out of print.

Launius, Roger D., and J. D. Hunley, compilers. *An Annotated Bibliography of the Apollo Program.* Monographs in Aerospace History, No. 2, 1994.

Launius, Roger D. *Apollo: A Retrospective Analysis.* Monographs in Aerospace History, No. 3, 1994.

Hansen, James R. *Enchanted Rendezvous: John C. Houbolt and the Genesis of the Lunar-Orbit Rendezvous Concept.* Monographs in Aerospace History, No. 4, 1995.

Gorn, Michael H. *Hugh L. Dryden's Career in Aviation and Space.* Monographs in Aerospace History, No. 5, 1996.

Powers, Sheryll Goecke. *Women in Flight Research at NASA Dryden Flight Research Center from 1946 to 1995.* Monographs in Aerospace History, No. 6, 1997.

Portree, David S. F., and Robert C. Trevino. *Walking to Olympus: An EVA Chronology.* Monographs in Aerospace History, No. 7, 1997.

Logsdon, John M., moderator. *Legislative Origins of the National Aeronautics and Space Act of 1958: Proceedings of an Oral History Workshop.* Monographs in Aerospace History, No. 8, 1998.

Rumerman, Judy A., compiler. *U.S. Human Spaceflight, A Record of Achievement 1961–1998.* Monographs in Aerospace History, No. 9, 1998.

Portree, David S. F. *NASA's Origins and the Dawn of the Space Age.* Monographs in Aerospace History, No. 10, 1998.

Logsdon, John M. *Together in Orbit: The Origins of International Cooperation in the Space Station.* Monographs in Aerospace History, No. 11, 1998.

Phillips, W. Hewitt. *Journey in Aeronautical Research: A Career at NASA Langley Research Center.* Monographs in Aerospace History, No. 12, 1998.

Braslow, Albert L. *A History of Suction-Type Laminar-Flow Control with Emphasis on Flight Research.* Monographs in Aerospace History, No. 13, 1999.

Logsdon, John M., moderator. *Managing the Moon Program: Lessons Learned From Apollo*. Monographs in Aerospace History, No. 14, 1999.

Perminov, V. G. *The Difficult Road to Mars: A Brief History of Mars Exploration in the Soviet Union*. Monographs in Aerospace History, No. 15, 1999.

Tucker, Tom. *Touchdown: The Development of Propulsion Controlled Aircraft at NASA Dryden*. Monographs in Aerospace History, No. 16, 1999.

Maisel, Martin, Demo J. Giulanetti, and Daniel C. Dugan, *The History of the XV-15 Tilt Rotor Research Aircraft: From Concept to Flight*. Monographs in Aerospace History, No. 17, 2000 (NASA SP-2000-4517).

Jenkins, Dennis R. *Hypersonics Before the Shuttle: A Concise History of the X-15 Research Airplane*. Monographs in Aerospace History, No. 18, 2000 (NASA SP-2000-4518).

Chambers, Joseph R. *Partners in Freedom: Contributions of the Langley Research Center to U.S. Military Aircraft of the 1990s*. Monographs in Aerospace History, No. 19, 2000 (NASA SP-2000-4519).

Waltman, Gene L. *Black Magic and Gremlins: Analog Flight Simulations at NASA's Flight Research Center*. Monographs in Aerospace History, No. 20, 2000 (NASA SP-2000-4520).

Portree, David S. F. *Humans to Mars: Fifty Years of Mission Planning, 1950–2000*. Monographs in Aerospace History, No. 21, 2001 (NASA SP-2001-4521).

Thompson, Milton O., with J. D. Hunley. *Flight Research: Problems Encountered and What They Should Teach Us*. Monographs in Aerospace History, No. 22, 2001 (NASA SP-2001-4522).

Tucker, Tom. *The Eclipse Project*. Monographs in Aerospace History, No. 23, 2001 (NASA SP-2001-4523).

Siddiqi, Asif A. *Deep Space Chronicle: A Chronology of Deep Space and Planetary Probes, 1958–2000*. Monographs in Aerospace History, No. 24, 2002 (NASA SP-2002-4524).

Merlin, Peter W. *Mach 3+: NASA/USAF YF-12 Flight Research, 1969–1979*. Monographs in Aerospace History, No. 25, 2001 (NASA SP-2001-4525).

Anderson, Seth B. *Memoirs of an Aeronautical Engineer: Flight Tests at Ames Research Center: 1940–1970*. Monographs in Aerospace History, No. 26, 2002 (NASA SP-2002-4526).

Renstrom, Arthur G. *Wilbur and Orville Wright: A Bibliography Commemorating the One-Hundredth Anniversary of the First Powered Flight on December 17, 1903*. Monographs in Aerospace History, No. 27, 2002 (NASA SP-2002-4527).

Chambers, Joseph R. *Concept to Reality: Contributions of the NASA Langley Research Center to U.S. Civil Aircraft of the 1990s*. Monographs in Aerospace History, No. 29, 2003 (NASA SP-2003-4529).

Peebles, Curtis, editor. *The Spoken Word: Recollections of Dryden History, The Early Years*. Monographs in Aerospace History, No. 30, 2003 (NASA SP-2003-4530).

Jenkins, Dennis R., Tony Landis, and Jay Miller. *American X-Vehicles: An Inventory—X-1 to X-50*. Monographs in Aerospace History, No. 31, 2003 (NASA SP-2003-4531).

Renstrom, Arthur G. *Wilbur and Orville Wright Chronology*. Monographs in Aerospace History, No. 32, 2003 (NASA SP-2003-4532).

Bowles, Mark D., and Robert S. Arrighi. *NASA's Nuclear Frontier: The Plum Brook Reactor Facility, 1941–2002*. Monographs in Aerospace History, No. 33, 2004 (NASA SP-2004-4533).

McCurdy, Howard E. *Low-Cost Innovation in Spaceflight*. Monographs in Aerospace History, No. 36, 2005 (NASA SP-2005-4536).

Seamans, Robert C. *Project Apollo: The Tough Decisions*. Monographs in Aerospace History, No. 37, 2005 (NASA SP-2005-4537).

Index

Advanced Earth Observing Satellite (ADEOS), 27, 41–42
Airborne Antarctic Ozone Experiment (AAOE), 19–20, 22, 31; Estes Park, 22
Airborne Arctic Stratospheric Expeditions (AASE) First, 23, 26, Second, 30
Albritton, Dan, 17–18, 19, 20, 21, 33, 37
Ames Research Center, 9
Anderson, Clinton, 3
Anderson, James, 15, 20, 22, 30, 31, 34
Anderson, Stephen, 2
Antarctica, Ozone hole, 14–15, 17–19, 20, 21, 22, 26, 29, 30, 33, 34, 38, 39, 42, 43
Arctic, Ozone hole, 26, 28, 30–34, 42, 43, 48
Asrar, Ghassem, 44
Association for the Advancement of Science (AAAS), 37
Atlantis, 38
Atomic Energy Commission, 7
Atmospheric Laboratory for Applications and Science (ATLAS), 36
Aura, 36, 43, 44

Baliunas, Sallie, 39
Beggs, James, 10–11
Benedick, Richard, 2
British Antarctic Survey, 14
Bromley, D. Allan, 26, 31
Bush, George H.W., 25, 26, 27, 29, 31–32, 35, 37–38
Bush, George W., 43

California Institute of Technology (Caltech) 9, 11, 15
Carter, Jimmy, 10, 11
Center for Climate Systems Research, 42
Challenger, 20, 25
Chemical Manufacturers Association, 18, 19, 26
Christie, Maureen, 2
Clean Air Act, 32; Amendments, 9
Climate Impact Assessment Program (CIAP), 3, 5, 7

Clinton, Bill, 36, 38, 39
Columbia University, 42
Committee on Earth Sciences (CES), 25
Condon, Estelle, 20, 26
Congress, 3, 5, 17, 19, 22, 25, 28, 35, 37
    Assigning ozone mandate, 7–9
    House Committee on Science, 38
    Republican revolution, 38, 39; Senate, 42
    Senate Committee on Aeronautical and Space Sciences, 3, 8
    Senate Subcommittee on Science, Technology, and Space, 26
Cook, Liz, 31

DC–8, 20, 30
Department of Commerce, 8
Department of Defense, 35
Department of Energy, 7
Department of Transportation (DOT) 3, 5, 7
Department of State, 12, 19, 21
Discovery 29, 36, 42
Doolittle, John, 39
Dukakis, Michael, 25
Dupont, 9, 22

Earth Day, 1
Earth Defense Fund, 37
Earth in the Balance, 36
Earth Observing System (EOS), 1, 25, 26, 28, 35–36, 37, 38, 39, 41, 43, 48
Earth Radiation Budget Satellite, 43
Earth System Sciences Committee, 11
Edelson, Burt, 10–11
Environmental Protection Agency (EPA), 10, 13, 15, 18, 21, 22, 31, 38
ER–2, 13, 19–20, 22, 30, 41
European Community, 26
European Space Agency, Earth Remote Sensing Satellite, 43, Envisat, 44

Farman, Joseph, 14–15, 18
Farmer, Crofton, 15
Federal Aviation Administration (FAA), 12

Federal Council on Science and Technology (FCST), 7
Fisk, Len, 29, 35
Fletcher, James, 3, 5, 9, 20
Ford, Gerald, 7, 9
Friends of the Earth, 31

Geneva, Switzerland, 18
Geophysical Research Letters, 18
George C. Marshall Institute, 39
Georgia Institute of Technology, 44
Goddard Space Flight Center, 9, 14, 15, 23, 28, 29, 44
Goddard Institute for Space Studies, 42
Goldin, Dan, 35–36
Gore, Al, 26, 31–32, 33, 36, 38
Graham, William, 18
Greenpeace, 31

Hampton, University, 44
Hansen, James, 25
Harris, Robert, 39
Harvard University, 15, 20, 22, 30, 34, 48
Heath, Donald, 10, 19, 22
Hilsenrath, Ernest, 38
Hodel, Donald, 19
Hofman, David, 33

Interagency Committee on Atmospheric Sciences (ICAS), 7

Jet Propulsion Laboratory (JPL), 9, 11, 15, 32
Johnson Space Center, 5, 9
Johnston, Harold, 3, 11

Kaye, Jack, 12, 15, 17
Kerr, Richard, 42
King, James, 9
Krueger, Arlin, 32
Kurylo, Michael, 30, 31, 34, 37, 41, 43

Landsat, 1
Langley Research Center, 9
Liftin, Karen, 12
Lohr, Karen, 31
Low, George, 7

McDonald, James, 3
McNeal, Joe, 30
Massachusetts Institute of Technology (MIT) SST study, 3
McPeters, Richard, 44
Meteor–3 satellite, 27, 29
Mission to Planet Earth, 18, 20, 25, 26, 28, 29, 35, 36, 38, 39, 43
Molina, Mario, 5, 11
Montreal Protocol on Substances That Deplete the Ozone Layer, 1, 20–21, 22–23, 27, 31, 32, 36, 37, 38, 39, 44, 47, 48
Morrison, Micah, 33
Moss, Frank, E., 8

National Aeronautics and Space Administration (NASA):
    Airborne Antarctic Ozone Experiment (AAOE), 19–20, 22, 31
    Airborne Arctic Stratospheric Expeditions (AASE) First, 23, 26; Second, 30
    Ames Research Center, 9, 20
    Atmospheric Laboratory for Applications and Science (ATLAS), 36
    Earth Observing System (EOS), 25, 26, 28, 35–36, 37, 38, 39, 41, 43
    Earth Sciences Enterprise, 43
    Earth System Sciences Committee, 11
    Goddard Space Flight Center, 9, 14, 15, 23, 28, 29, 44
    Goddard Institute for Space Studies, 42
    Headquarters, 5, 29, 31, 44
    Jet Propulsion Laboratory, 9, 11, 15, 32
    Johnson Space Center, 5, 9
    Langley Research Center, 9
    Mission to Planet Earth, 18, 20, 25, 26, 28, 29, 35, 36, 38, 39, 43
    National Ozone Expedition (NOZE), 18, 20
    Office of Space Science and Applications (OSSA), 9, 10, 29, 35
    Space Shuttle, 5, 20, 25, 29, 36, 38, 42
    Space Station, 10
    Upper Atmosphere Research Office (UARO), 9
    Upper Atmosphere Research Program (UARP), 10, 11, 17, 20, 26, 34, 41

Upper Atmosphere Research Satellite (UARS), 10, 11, 25, 28, 29–30, 32, 35, 36, 37, 38, 39, 41, 42–44
U.S. Global Change Research Program, 25, 26, 38
National Oceanic and Atmospheric Administration (NOAA), 7, 12, 17, 33, 37, 38, 42, 44
    Airborne Antarctic Ozone Experiment (AAOE), 19–20
    Airborne Arctic Stratospheric Expeditions (AASE) First, 23, 26; Second, 30
    Aeronomy Laboratory, 15, 17, 20, 22, 34
    Climate Monitoring and Diagnostics Laboratory, 33
    Environmental Technology Laboratory, 33
    National Ozone Expedition (NOZE), 18
    U.S. Global Change Research Program, 25, 26
National Ozone Expedition (NOZE), 18, 20
National Research Council, 3, 9, 39
National Science Foundation (NSF), 7, 11, 18, 19, 23, 47
    Airborne Arctic Stratospheric Expeditions (AASE) First, 23, 26, Second, 30
    National Center for Atmospheric Research (NCAR), 30, 33
    U.S. Global Change Research Program, 25, 26
Nature, 5, 15, 37
Naugle, John, 7
Newchurch, Michael, 45
Newman, Paul, 44
New York Times, 3, 31, 37
Nimbus 7, 9, 13, 25, 27, 29, 41
Nixon, Richard M., 3, 5

Office of Science and Technology Policy (OSTP), 37, 39
Office of Space Science and Applications (OSSA), 9, 10, 29, 35
Oppenheimer, Michael, 37
Ozone Trends Panel, 19, 20, 22, 25, 30

Parson, Edward 2, 48
Present State of Knowledge of the Upper Atmosphere: An Assessment Report, 12, 17
Project Habitat, 10

Reagan, Ronald, 10, 11, 12, 18, 19, 22, 25
Reber, Carl, 29
Reilly, Bill, 31
Ride, Sally, 20, 25
Roland, F. Sherwood, 5, 11

Sarma, K. M., 2
Satellite Pour l'Observation de la Terre (SPOT 4), 43
Schmeltekopf, Art, 15, 19
Schoeberl, Mark, 28, 38, 44
Shapiro, Melvyn, 33, 34
Shultz, George, 19
Science 3, 42
Senate Committee on Aeronautical and Space Sciences, 3, 8
Senate Subcommittee on Science, Technology, and Space, 26
Singer, Fred, 39
Solomon, Susan, 18, 38
Soyuz rocket, 42
Space Shuttle, 5, 20, 25, 29, 36, 38, 42; Space Shuttle's environmental impacts, 5, 7
Space Station, 10
Steitz, David, 43–44
Stever, H. Guyford, 7
Stolarski, Richard, 23
Sununu, John, 31
Supersonic Transport (SST), 3, 5

Thomas, Lee, 13, 15, 18
Tilford, Shelby, 13, 20, 29, 30, 35
Time, 31, 32
Todd, Ed, 7
Tolba, Mostafa, 12
Total Ozone Mapping Spectrometer (TOMS), 13, 14, 25, 27, 29, 32; Earth Probe, 41–43, 44; QuikTOMS, 44
Truly, Richard, 26, 35
TRW, 35
Tuck, A.F., 20

United Kingdom Meteorological Office, 19

United Nations, 2, 32
　　Conference on the Peaceful Uses of Outer Space, 10
　　Environmental Program (UNEP), 12
　　UNEP/WMO Scientific Assessment of Ozone Depletion, 44
　　Intergovernmental Panel on Climate Change (IPCC), 25
University of Arizona, 3
University of Alabama at Huntsville, 44, 45
University of California at Berkeley, 3, 11
University of Maryland, 11
University of Virginia, 39
University of Wisconsin, 44
Upper Atmosphere Research Office (UARO), 9
Upper Atmosphere Research Program (UARP), 10, 11, 17, 20, 26, 34, 41
Upper Atmosphere Research Satellite (UARS), 10, 11, 25, 28, 29–30, 32, 35, 36, 37, 38, 39, 41, 42–44
U.S. Air Force Midcourse Space Experiment, 43
U.S. Global Change Research Program (USGCRP), 25, 26, 38
U.S. Scout class booster, 27

Vienna Convention for the Protection of the Ozone Layer, 13, 18

Walker, Robert, 38
Wall Street Journal, 34
Waters, Joe 32
Watson, Robert, 15, 30, 36, 48
　　Airborne Antarctic Ozone Experiment (AAOE), 19–20
　　Albritton alliance, 17, 33, 37
　　Arctic expedition, 26
　　Background, 11
　　Managerial strategy, 11, 19, 48
　　Montreal Protocol, 20–21
　　National Ozone Expedition, 18
　　Office of Science and Technology Policy (OSTP), 37, 39
　　Ozone Trends Panel, 19, 22
　　Policy entrepreneur, 11–13
　　WMO/NASA Assessment, 12, 17

UNEP, 12
Washington Post, 31, 32
Washington Times, 34, 39
World Meteorological Organization (WMO), 12, 44
WMO/NASA assessment, 12, 17